THE SCIENCE OF SUPERHEROES AND SPACE WARRIORS

Lightsabers, Batmobiles, Kryptonite, and More!

sourcebooks

Published by Sourcebooks, Inc.
P.O. Box 4410, Naperville, Illinois 60567-4410
(630) 961-3900
Fax: (630) 961-2168
www.sourcebooks.com

Library of Congress Cataloging-in-Publication Data

The science of superheroes and space warriors : lightsabers, Batmobiles, kryptonite, and more!
      pages cm
  Includes bibliographical references.
  1. Science in popular culture. 2. Physics--Popular works. 3. Comic books, strips, etc.
  QC24.5.S45 2014
  500--dc23

2014025241

Printed and bound in the United States of America.
VP 10 9 8 7 6 5 4 3 2 1

CONTENTS

WHY WE LOVE SUPERHEROES AND SUPERPOWERS

Since the dawn of time, or at least the dawn of cartoons and comic books, kids have imagined themselves with superpowers. "Wonder Twin powers, activate!" As kids grow older, their dreams of flying or talking to animals or being invisible largely fade away. Yet our general fascination with the idea of these extraordinary abilities has never dwindled. In ancient cultures, mythical beings often possessed special powers that set them apart from mortal men. If Zeus zinging a lightning bolt in displeasure wasn't a display of a superpower, then what *was*?

It's human nature to pursue the most impossible aspects of the folklore (or cartoons) we've created and attempt to make them realities. In fact, that's one of the awesome ways in which art and imagination drive science. Creative minds dream up ingenious ideas like invisibility cloaks, and scientific minds follow through by inventing them.

That's no joke! As you'll find out later on, people are hard at work inventing invisibility cloaks and lots of other gizmos and gadgets to give us ordinary folk real versions of superpowers. In this book, we, the staff at HowStuffWorks, decided to tackle entire super-legends in a single bound to discover what exactly it takes to be super.

First, we'll look at the science behind your favorite superheroes and supervillains and their super-cool devices and weapons, from the Batmobile and warp speed to lightsabers, death stars, and kryptonite. Then we'll delve into the latest and greatest inventions—from liquid body armor to invisibility cloaks, replicants, and limb regeneration. We even pit Superman against some other super-warriors—including Harry Potter, Sith Lords, and even Chuck Norris—to see who would win.

Throughout the book, we've packed in lots of fun elements, including trivia tidbits and sidebars, so you can maximize your super know-how. We've also thrown in a few quizzes throughout to determine what you've learned so far. After all, knowledge is its own superpower.

Obviously, some of the superpowers explored here are still light-years away from reality, but you might be pleasantly surprised at the powers science has already bestowed upon us mere mortals. Onward and upward!

SCIENCE IS SUPER ALREADY: TOP FIVE SCIENCE-BORNE SUPERPOWERS

Before we get into the science behind our favorite superheroes and supervillains and how their superpowers might work in real life, let's take a look at five awesome superpowers that are rapidly becoming everyday realities!

1. Seeing through Walls

"You make a better wall than a window." Remember that old saying? It was something you'd spout off to someone while he or she was obstructing your view of whatever you were trying to see. But several optics companies are rendering the expression obsolete.

For example, Camero's Xaver 800 product uses microwave radar to penetrate walls and project 3-D imaging of whatever is hiding behind those walls. According to the company, ordinary drywall, clay brick, cinder block, and even rebar-reinforced concrete structures are no match for the Xaver 800. However, just as Superman could be stymied by kryptonite, the device can't see through solid, continuous metal. When in use, the device takes up an area of about 33 by 33 by 6 inches (84 by 84 by 15 centimeters) and weighs almost 33 pounds (15 kilograms).

Other companies are also getting into the X-ray-vision business. Physical Optics Corporation offers a hand-held X-ray scanner called LEXID that can reveal contraband hidden behind walls, in cars, and in sealed containers. ThruVision's T5000 people screener can detect concealed weapons without revealing intimate details about a person's body.

These companies primarily design their products for use in law enforcement, military, fire and rescue, and security applications. You can see how these gadgets might come in handy if, for instance, someone's been taken hostage and negotiators are working to end the standoff safely. That is certainly an activity befitting a superhero.

Next we'll find out how to elude evil forces.

2. Defying Gravity

Although most of us are bound by the force of gravity, evolution has helped some critters—like the gecko (and Spider-Man)—to circumvent that force. To ensure their survival, these animals developed the ability to fasten themselves effortlessly to most walls and ceilings without leaving behind a sticky residue.

Scientists have been working to develop a synthetic adhesive that mimics the gecko's special dry stickiness for years, and such an invention is very close to becoming a real-ity, thanks to a new method of using plastic to create similar microscopic structures.

In all likelihood, space will be the first destination for the new dry adhesives. Currently, the idea is that the tech-nology will assist robots in the exploration of the final fron-tier. However, it's entirely plausible that the same principles could eventually be applied to replace the everyday adhe-sives to which people have grown accustomed. Good-bye, sticky tape. Hello, sticky feet.

Not that you should say good-bye to your beloved tape, just yet. The design hasn't been perfected yet, and one of the problems currently hampering development is that, in some cases, the adhesive is too sticky, making it difficult to remove once applied. So this superpower is still in the works.

⚡3. Superhuman Strength

The human body can be conditioned to endure extreme environments, but some situations call for strength above and beyond our natural abilities. On the battlefield, soldiers need to carry heavy loads over extended periods and through harsh ter-rains. So to push the limits of physical exertion, scientists have developed a way to let technology bear some of that burden.

Berkeley Bionics and Lockheed Martin's Human Universal Load Carrier (HULC) exoskeletons are an example of such innovation. Through its titanium legs, HULC does for human skeletons on the outside what adamantium did for Wolverine's skeleton on the inside, turning its wearer into one tough customer.

The carrier's titanium legs are mounted to a backpack-like frame, which houses a power unit and a small on-board microcomputer. One of HULC's most impressive features is that it does not require a joystick or manual control mechanism. Not unlike Iron Man's famous suit, the device can sense the operator's intended movements and react accordingly.

To minimize the strain upon the user, the HULC design uses hydraulics, which make the deep squats and heavy upper-body lifting seem as easy as the blink of an eye. Right now, its applications are specifically military-minded, but defense giant Lockheed Martin is exploring options for its use in industrial and medical capacities.

⇌ 4. Stopping Bullets

It's unlikely that you'll ever be faster than a speeding bullet on your own. But if an invention could actually sense when an incoming round was on its way to meet you, well, that would be something. And if such a device could also compel you to dive out of harm's way, à la Neo in *The Matrix*, it would surely rank among the most awesome lab-developed superpowers.

IBM filed a patent for Bionic Body Armor, which would accomplish such tasks, in early 2009. The device would have induced a shock to the wearer that would, in turn, have caused him to make reflexive movements in the opposite direction from a threatening projectile. Unfortunately, IBM pulled the patent in February of the same year.

Never fear, though. As we'll see later on, smart armor is still being developed in the form of liquid body armor. One version uses magnetorheological fluids, which thicken when exposed to a magnetic field, and another uses shear-thickening fluids, which harden when agitated or struck forcefully by an object.

As these forms of armor become more refined, they will certainly change the landscape of war. And while they haven't been perfected yet, their value in terms of lives saved and casualties prevented will be immeasurable. That's why smart armor will grant users one of the most awesome superpowers of all.

5. Manipulating the Weather

Part of what separates human beings from other animals is our propensity to change our environment to accommodate our needs. But some things elude our direct control, and weather is one of them, unless you're Storm from the X–Men. And who doesn't want to be? After all, how cool would it be to summon a sunny day at a moment's notice? Or generate a blizzard on a whim when you feel like having a snow day?

If only we had Storm's superpower of weather manipulation, we might even be able to work out global warming. Think of it. Not only could we cool down the planet, but we could also end droughts, and sports fans would never be forced to endure another frustrating rain delay.

Although it's a far cry from Storm's excellent mutant abilities, scientists do have one technique up their sleeve for manipulating the weather: cloud seeding.

Cloud seeding can trace its roots back to the United States in the 1940s. Today, it's used to increase precipitation, disperse clouds and fog, and suppress hail. Depending on the environment and objectives, chemicals can be shot up from the ground or released in midair. There is a variety of cloud-seeding agents available including silver iodide, salt, and ammonium nitrate.

The Weather Modification Association issued a statement in July 2009 saying that cloud seeding with silver iodide doesn't harm the environment, and studies have shown the process to increase precipitation by up to 30 percent in some cases. With the ability to manipulate the weather in certain situations, scientists are gaining on superheroes, but they still have a long way to fly.

HOW COMIC BOOKS WORK

Blam! Zap! Pow! These are more than page-bound sound effects; they're emblems of a unique form of human communication. They're a hallmark of— you guessed it—comic books.

When you think of comic books, you may conjure up images of geeky adolescent boys and superheroes in stretchy tights portrayed in flimsy little booklets. But there's way more to this game-changing, generation-spanning art form than that. What's the real story of comic books? What do comic-book fans look like today?

Comic books have a storied place in the history of human publishing and strong roots in American culture. In many ways, they're the apple pie and Fourth of July versions of American literature, full of iconic imagery, action, drama, and sometimes even artistry worthy of a gallery in New York's Museum of Modern Art.

COMIC BOOK DEFINITIONS

There are so many varieties of comic books and related types of work that it's worth reviewing what "comic book" and some related terms mean.

❋ A **comic book** blends drawn art—color, black-and-white, or both—with text to tell a story.

❋ Comic books are usually twenty to thirty pages long. Once they get longer, they're sometimes called **graphic novels**.

❋ Stylistically, comic books are easy to spot. Each page of the books is segmented into **panels** (or frames), which have borders that separate them from other panels.

❋ Individual panels contain one part of a story (perhaps dialogue between characters) or a character's inner thoughts (represented by **speech** and **thought balloons**) that lead into the next panel.

❋ Panels are routinely separated by blank areas called **gutters**. Artists lay out each page so that panels logically flow, guiding the readers' eyes so that they absorb the story in a sequential manner. For this reason, comic books are often called **sequential art**—a type of graphic storytelling.

It's not uncommon for people to scoff at anyone who uses the word "art" in describing comics. Comics have a long history that's been linked to low-brow, cheap entertainment. But comics these days aren't just longer versions of the Sunday funnies. They and their artistic cousins, graphic novels and superhero movies, saturate our society.

You'll find comic strips and comic books at newsstands everywhere, and graphic novels and comics in strip-mall bookstores and geek boutiques all across the land. Comic-based movies now routinely show up in theaters and draw huge crowds. It's not a surprise that these flicks are popular. Comic books have inspired a fanatical subculture, one that features its own lingo, slang, and inside jokes, as any fan of Kevin Smith's films can attest.

Comic books and graphic novels span as many genres and subjects, both frivolous and mature, as traditional novels, and they've created some of the most revered and recognized figures: Superman, Batman, Spider-Man, Hulk, Xena: Warrior Princess, Carmen Sandiego, and many, many more.

Comics are a global phenomenon, but in this section, we'll focus on the American evolution of comic art, with nods to notable international examples. Keep reading, and you'll see how comic books first came to life, how they're more popular than ever, and how their creators keep finding new audiences to enthrall.

⇌ Dusting Off the (History of) Comic Books

Ask three experts to name the very first comic book, and you'll probably get three different answers, depending on the exact semantics they use to define the medium.

However, it was in 1842 that a book complete with paneled drawings and captions made its first appearance in America. It was *The Adventures of Obadiah Oldbuck*, by Rodolphe Töpffer, a Swiss teacher and artist. Many historians consider *Obadiah* to be the first comic book. After Töpffer's breakthrough, comics spread slowly but surely in print media.

Modern comic books grew out of comic strips, which were almost always humorous. (Thus, the word "comic" stuck.) Those short narratives with just a few panels, such as *The Yellow Kid*, graced the pages of newspapers beginning in 1895. *The Yellow Kid* used speech balloons, a convention still used in modern comics.

Comic strips were just one of the phases in the evolution of comic books. From the turn of the century to the 1930s, publishers cranked out hundreds of different strips featuring still-famous characters such as Dick Tracy, Popeye, and Little Orphan Annie. And these strips didn't just appear in the Sunday funnies. Other forms of comics appeared on newsstands and in gas stations in the form of cartoon books and comics used as advertising gimmicks.

In 1933, the modern comic began to take shape in the form of *Funnies on Parade*. Eastern Color Printing reprinted this short collection of comic strips on tabloid-sized pages, which simply meant the publication was a smaller size than newspapers of the time. *Funnies* wasn't for sale; rather, Procter & Gamble gave the books away as a part of what turned out to be a very successful corporate promotion.

Eastern Color decided to capitalize on its momentum by actually selling a book of comics. These were still reprints of funny Sunday comic strips, and they sold quite well on newsstands, opening the door for the next big jump in the comic book metamorphosis.

The epic adventure of comic books had just begun. Let's explore how newfangled comic books and their lead characters not only revolutionized comic art, but perhaps (with only a smidgen of hyperbole) changed the course of humanity.

COMICS ACROSS THE POND

Europe has its own rich comics history. Just as a cheeseburger in France is distinctive from one in the United States, the art of comics looks very different depending on which side of the pond it came from. Landmark titles like *TinTin* (Belgium), *Asterix and Obelix* (France), *Tex Willer* (Italy), *Agent 372*

(Netherlands), *The Beano* (England), and even *The Smurfs* (Belgium) originated in Europe and have their own stylistic tendencies and themes.

⚡ Drawn in Steel: How Comic Books As We Know Them Came to Be

In 1935, DC Comics printed the first comic book filled with mostly new material instead of comic strip reprints. The title was *New Fun Comics No. 1*, and it introduced the American public to brand-new characters and story lines that spanned several issues as parts of series. Ninety years after *Obadiah*, comics were finally becoming a true cultural force.

The sixth issue of *New Fun* featured work from Jerry Siegel and Joe Shuster, the two men who gave life to a character that changed the face of comic books—and, in some ways, American culture itself.

That character's name? Superman. Yep, you know the one: faster than a speeding bullet, more powerful than a locomotive, the Man of Steel. Comic-book legend and world-famous American icon.

In June 1938, Superman debuted in *Action Comics No. 1*. This was the first superhero comic book, and Superman's success gave rise to all manner of superheroes, from Batman and

Captain America to the Fantastic Four and Wonder Woman. That first appearance of Superman is so legendary that a copy of *Action Comics No. 1* can sell for more than $3 million.

Superman wasn't the only legend born in those days. DC Comics also published its *Detective Comics* series around the same time, in 1937. This publication introduced Batman to the world. The *Detective* series is still going strong today, with more than 800 issues, and it stands as the longest-running comic book title ever.

During World War II, superhero-themed comic books were enormously popular. In some of the most popular titles, iconic superheroes battled (and always defeated) the forces of evil in ongoing series that lasted for multiple issues published over many months and years. Other, non-hero characters were hugely popular, too. Walt Disney's lineup, in particular, scored big with characters such as Mickey Mouse and Donald Duck.

Once the war ended, many among the plethora of superheroes lost their luster. To recapture and enlarge their readership, comics publishers introduced more diverse subjects, featuring science fiction, drama, animal, Western, crime, and horror comic books. Those last two categories, in particular, led to another seismic shift in the comics landscape. In fact, as we'll find out in the next section, a single (non-comic) book completely warped comic-book publishing and disenchanted readers everywhere—and it took a comic-book revolt to save the medium from itself.

⚡ Kryptonite to the Authority

In 1954, psychiatrist and author Frederic Wertham published a book titled *Seduction of the Innocent*, which scapegoated comic books, and horror comics in particular, as a cause of adolescent depravity and misbehavior. Parents everywhere were alarmed, and a cultural crusade to clean up comics began.

To save their industry, comics publishers created a self-censoring Comics Code Authority (CCA), which set standards for comics content. Mostly, they whitewashed their comics to make them completely inoffensive, politically correct, and nonthreatening to children and institutions of the time. For example, no title could use the words "terror" or "horror." Other prohibited characters and themes included vampires, zombies, torture, and werewolves.

Mainstream comics became banal ghosts of themselves. And, thanks to the prominent Comics Code Authority label, parents felt comforted that the materials their children were reading wouldn't eventually turn them into serial murderers.

The CCA may or may not have prevented comics from inspiring legions of Jeffrey Dahmers, but it did help spark an underground comics movement. In the 1960s, independent publishers and authors, unbound by the CCA rules, began making comics featuring every sort of subject that mainstream comics couldn't—sex, drugs, politics, and both visual and written obscenities of every kind. These alternative, underground comics were often called comix. Comix spawned their own superstars, including Fritz the Cat, Trashman, and Wonder Wart-Hog.

The unbridled creativity of comix heralded another new era of the comic-book medium. Unburdened by sales quotas and censorship, artists crafted comics with more sophisticated story lines and themes, often trading campy superhero fluff for mature, literary writing styles matched by equally advanced artwork.

The lowly comic book, which had been vilified by the press, derided by lovers of "true" art, and ripped from the hands of countless students by ticked-off teachers, suddenly began gaining respectability.

CRACKING THE CODE

In 2011, publishing giant DC Comics decided to officially end its adherence to the Comics Code Authority, ending more than fifty years of self-censorship. Marvel Comics, another major player, had already gone that route years before.

⚡A Medium's Metamorphosis

In the 1970s and 1980s, longer-form comics—sometimes serious, sometimes funny—became more common and were sometimes labeled graphic novels. These so-called "novels" featured complicated subjects such as morality and philosophy, and they introduced conflicted characters and damaged, fallible superheroes.

In some particularly notable works, it became clear that the power and scope of comic art had become a true cultural force. *Watchmen*, *Maus*, and *The Dark Knight Returns* put on display the kind of nuanced, tour-de-force works that were possible when scholarly writers matched their intellect with equally talented illustrators. Since these titles were released, comic book and graphic novel creators have continued to explore and find success in every genre.

WHO SUPPLIES OUR SUPERHEROES?

Independent publishers do sell many comic books. But these days, two long-standing rivals, DC Comics and Marvel, account for about eight of every ten comic books sold in the United States. Their hit titles, which often benefit from movie adaptations, include *Superman*, *X-Men*, *The Dark Knight*, and *Iron Man*, among many others.

While the U.S. sequential art medium was experiencing its developmental renaissance, Japan saw a similar rise in comic art. After World War II, especially, the Japanese fell in love with comics and began publishing them en masse. Japanese comics are generally referred to as "manga." Manga is a massive cultural spectacle, more popular in Japan than comic books are in America. Manga creators make books that target and grab every section of society and gender, from young children and teens to adults.

Perhaps unsurprisingly, manga is also growing in popularity in the United States. Sometimes manga titles, such as *Fake* and *Death Note*, even outsell those made in America. Manga often uses simple black-and-white drawings, with only occasional highlights of color, and features stories that appeal to girls as well as to boys. In fact, when it comes to manga, girls are the driving market force, and most titles are geared toward their tastes.

These days, in both Japan and America, comic creativity is still high, drawing on the powers of top-notch writers and artists. But how do these talented artists make their characters and stories leap from the page?

Artful Comic Architects

Typically, comic books aren't solitary creative pursuits. Rather, teams made up of a writer, a penciler, an inker, a colorist, and a letterer often combine talents to make for the most compelling books.

* **Writers** often conceptualize the plot of a comic book and write the dialogue and captions. They imagine the way the words and art play off each other to create an impactful, cohesive story. In other situations, the writers collaborate extensively with the other artists in the group.

* **Pencilers** draw the rough initial panels and characters that make up a particular comic universe. These are usually just sketches, but they're the basis for all of the drawn art that follows.

* Some comics teams also have **inkers**, who add to and enliven the work of pencilers. Using complex shading schemes, they use their creative flair to add drama and motion to each panel.

* **Colorists** add color to the black-and-white drawings in each panel. Colorists choose a specific color scheme, or palette, that matches the mood and tone of the story. Dark hues might add drama, while brighter tones may add a sense of joy or wonder.

* **Letterers** are tasked with adding life to captions and thought and speech balloons. Bold fonts convey emphasis. Squiggly letters can convey uncertainty or fear.

Laura Martin, a colorist for Marvel Comics, says that quite a few artists take on the challenge of doing all of the work for a book by themselves. But mainstream comics typically arrive at a rate of an issue per month, so time restrictions make the team approach much more efficient. Most pencilers and inkers crank out roughly one page per day, so by the end of the month, a twenty- to thirty-page comic book is ready to go.

So how do all these people get into this business?

THE FREE COMICS TRADITION

Every year, on the first Saturday of May, a very special event is staged, just for boys and girls who have been extra nice. Actually, Free Comic Book Day is open to naughty people, too, as comic shops everywhere give away certain titles for free to fans of all sorts. They do so to thank loyal readers and attract new audiences to their stores.

It's a Comics Life for Me

Mainstream comics publishing is demanding, deadline-driven work. Brian Stelfreeze, who has drawn dozens of cover illustrations for DC Comics and Marvel, among other clients, says that in this environment, teamwork is paramount, although he found the idea of collaboration a little strange at first.

"I started my career as a commercial illustrator, so the process of an assignment divided among three artists seemed quite foreign to me," he says. Now, though, he's sold on the teamwork approach. "Comics depend on the synergy of specialists and creating a work greater than the sum of its parts, but I find it often only as strong as its weakest link."

You may wonder how a world-class artist winds up in comic books. The road to comics success is not necessarily a straight one. Like many comic artists, Stelfreeze has a patchwork background. He started by drawing editorial cartoons in high school, went on to airbrush T-shirts in Myrtle Beach, South Carolina, and jettisoned art school for a job as a commercial illustrator. Then he fell back in love with comics.

"I had always wanted to draw comics as a kid, so I thought I'd try it just once," he says. "Fateful last words spoken by many an alcoholic, junkie, (and) vampire."

Laura Martin at Marvel got bitten by the comics bug while in college. She says, "I read comics as a child and in my early teens, but I'd largely dismissed them until I was in college. At the University of Central Florida, my major was graphic design and my minor was art history. While I was going to school there, I worked nights at Kinko's, where several of the employees were comics fans. Their enthusiasm got me excited about comics again."

Martin says that her coloring work is 99 percent digital, but that style and techniques vary depending on the artist. "Comic coloring is part of a collaboration. We're not creating new art ourselves. Our job is to enhance the black-and-white artwork drawn by a penciler and inker, which is drawn based on a story written by a writer."

That means individuals aren't in complete creative control—their work is dictated by the story's setting and mood, among other factors. Again, efficiency is key. "Many pros color over sixty pages per month. So we have techniques and styles in place to maximize productivity while not sacrificing storytelling or style," Martin says.

Reading this description of the comics workflow might make you think the whole process sounds vaguely cinematic. How do these artists' works wind up on the big screen?

Comics Take On the Silver Screen

Comics have enjoyed so much success in the past two decades that in some form or another, they've infiltrated all parts of our society. Hollywood, for instance, is awash with comic book adaptations.

From *Spider-Man* and *Watchmen* to *From Hell* and *Ghost*

World, comic books offer film producers plentiful material for on-screen action and suspense. Sometimes those films work spectacularly and translate into huge success at the box office.

The Dark Knight, an epic flick starring Batman and created by Warner Bros., grossed more than $500 million. *Iron Man* also scored big, with more than $300 million in gross ticket sales. All three *Spider-Man* movies grossed more than $300 million each. *Barb Wire*, on the other hand, was a spectacular flop, earning less than $4 million.

Clearly, audiences can connect with comics that are brought to life on the big screen. Stelfreeze says that when it comes to Hollywood hits, certain comics work better than others.

"I think comic-book movies face a difficult challenge because they are generally as good as the material they come from. The problem is that we consume movies but we interpret comics," he says.

Stelfreeze thinks comics are a more immersive and interactive medium, one in which readers fill in the moments between panels, in effect providing the quality of directing and acting.

"The comic book often provides less than 10 percent of the story itself, so the quality is consistent with the reader's imagination. Movies simply are what they are. I believe the last *Punisher* movie was about as good as your average

Punisher comic, but every reader has this fantasy Punisher in their heads that's much better than both."

Movies are hardly the last stop for comic books. In fact, comics are finding fame in many other parts of America's cultural dialogue.

⚡ The Imminence of Comics

It's fairly clear nowadays that comic books are anything but a passing fad. Their long history and current resurgence means comics have become cultural touchstones that make appearances just about everywhere.

Classrooms and libraries used to be no-fly zones for Superman and Captain Marvel. These days, teachers and librarians actually leverage comics to pry into the minds of unenthusiastic or fearful readers. Sometimes, students who struggle with the written word digest literary devices like symbolism, themes, and narrative better when there are pictures to help them along.

Comics are also useful for communicating nonfiction or biographical information in an engaging and relatable way. Politicians and celebrities are frequent subjects. For instance, Bluewater Productions, an independent comic publisher, has already produced comics and graphic novels about the lives of Donald Trump, Oprah Winfrey, Barack Obama, Justin Bieber, and Jon Stewart.

And, thanks to the Web, you no longer even have to visit a newsstand or comic-book shop to find comics. Web comics are a genre unto themselves, with all of the unlimited potential of the digital tools used to create them. In fact, many Web comic artists call the computer screen the infinite canvas, one that can be used to show the artistry of comics in a single panel or innumerable frames with imaginative layouts and limitless storytelling potential.

Perhaps the infinite canvas is an appropriate phrase for comics as a whole. They've morphed from simple, one-frame beginnings to novel-length epics, and then to the movie screen and the online world. They've triumphed over "funny book" and superhero stereotypes, survived government suppression, and even won over the very teachers who once banned them from English class. If all of this history is any indication, comic books will be here spinning their tales for a long time to come.

HEROES UNMASKED: HOW SECRET ARE SECRET IDENTITIES?

Now that we've taken a look at the evolution of comic books, let's move on to some of their most famous stars: superheroes and supervillains. What do you think of when you hear those two words? They probably conjure up images of anonymous masked, muscled men and women in tight uniforms or menacing armor—or something along those lines.

But these days, the anonymity of super-people (whether good or bad) is becoming extinct faster than Superman in a cell full of kryptonite. Well-hidden alter egos are few and far between. Between the big two (Marvel and DC), only five major heroes have yet to fully reveal their civilian personas—Superman, Spider-Man, Batman, Daredevil, and The Flash.

So when more than a dozen people know your "secret identity," is it really fair to call it secret? In this section, comic book pundit Matt Hunt reveals just how many characters are in on the superheroes' darkest secrets. Let's find an empty phone booth and examine which of the five unrevealed comic-book icons are most likely to lose their social camouflage before their superpowers.

THE FAST AND THE CURIOUS

COMIC BOOK ICON: The Flash

SECRET IDENTITY: Wally West, one-time Kid Flash and nephew of the Silver Age Flash, Barry Allen

ALLIES AWARE OF TRUE IDENTITY: Linda Park (wife) and the Justice League of America. (Current members of the JLA, including Aquaman, Batman, Green Lantern, Martian Manhunter, Plastic Man, Superman, and Wonder Woman, revealed their identities to each other.)

ENEMIES AWARE OF TRUE IDENTITY: None

TOTAL: 8

CHANCES OF BEING EXPOSED IN THE NEAR FUTURE: Minimal. The Flash's identity was once public knowledge, but after losing his unborn child to his arch-foe, Professor Zoom, Wally West agreed to accept an offer from the supernatural being known as the Spectre to make the entire world forget that he was The Flash. Only those closest to Wally, including his wife and members of the JLA, have eventually remembered his dual identity. With no villains quick enough to decipher the truth, The Flash is in the clear... for now.

SUPE'S ON

COMIC BOOK ICON: Superman

SECRET IDENTITY: Clark Kent, *Daily Planet* reporter

ALLIES AWARE OF CIVILIAN IDENTITY: The JLA (Aquaman, Batman, The Flash, Green Lantern, Martian Manhunter, Plastic Man, and Wonder Woman), Black Racer, Dr. Occult, the Eradicator, Green Arrow, Jon and Martha Kent (his foster parents), Lana Lang (ex-girlfriend), Lois Lane (wife), Lori Lemaris (ex-girlfriend), Nightwing, Phantom Stranger, The Spectre, Steel, Superboy, Supergirl, and Waverider

ENEMIES AWARE OF CIVILIAN IDENTITY: Bizarro, Brainiac, the Cyborg, Dominus, and Mr. Mxyzptlk

TOTAL BEINGS AWARE OF CIVILIAN IDENTITY: 28

CHANCES OF BEING EXPOSED IN THE NEAR FUTURE: Not likely. While most of those aware of his public persona are fellow heroes, the few villains privy to this knowledge are either currently out of commission (Dominus and the Cyborg are trapped in the phantom zone; Brainiac is stuck at the beginning of time) or could care less. (Bizarro and Mxyzptlk only care about the challenge of battling Superman.) Though it wouldn't hurt to be a little more hesitant to let his pals in on the secret, Supes should be ducking into phone booths for a long time.

DANGLING BY A THREAD

COMIC BOOK ICON: Spider-Man

SECRET IDENTITY: Peter Parker, *Daily Bugle* photographer

ALLIES AWARE OF HIS SECRET IDENTITY: Angel, Aunt May, the Black Cat, Bounty, Daredevil, Dr. Strange, Ka-Zar, Kaine (Peter Parker clone), Kraven, Madame Webb, Mary Jane Watson-Parker (wife), Nick Fury, the Sentry, Spindrifter, Thor, and Wolverine

ENEMIES AWARE OF SECRET IDENTITY: Chakra, Dr. Octopus, Green Goblin, Judas Traveller, Mendel Stromm, Mephisto, Nightmare, Puma, Stunner, Thanos, and Venom

TOTAL: 28

CHANCES OF BEING EXPOSED IN THE NEAR FUTURE: Even. With many of his knowledgeable enemies either comatose, amnesic, inactive, or too powerful to care, one would think Spider-Man is sitting pretty—right? Wrong. The only thing keeping Spidey's deadliest foe, the Green Goblin (a.k.a. Norman Osborn), from exposing his secret ID is the Goblin's insanity and his twisted sense of honor and respect.

On numerous occasions, Osborn has thrown monkey wrenches into Parker's personal life, including purchasing the newspaper where he works, kidnapping his newborn child, and even replacing his Aunt May with a genetically altered actress. After Osborn's public identity was revealed and he was sent to prison, exposing Parker might be Osborn's last act of vengeance.

THE BAT'S OUT OF THE BAG

COMIC BOOK ICON: Batman

SECRET IDENTITY: Billionaire playboy and Gotham City socialite Bruce Wayne

ALLIES AWARE OF TRUE IDENTITY: The JLA, Azrael, Batgirl, Sasha Bordeaux, Catwoman, Henri Ducard, Green Arrow, Carleton LeHah, Man-Bat, Nightwing, Nomoz, Oracle, the Outsiders, Alfred Pennyworth, Robin, Sentinel, Leslie Thompkins, Waverider, and Wildcat

ENEMIES AWARE OF TRUE IDENTITY: Bane, Bird, David Cain, Deathstroke, Hush, Lady Shiva, Nyssa, Ra's Al Ghul, the Riddler, Talia, Trogg, Ubu, and Zombie

TOTAL: 43

CHANCES OF BEING EXPOSED IN THE NEAR FUTURE: Highly likely. The villain Bane showed just how dangerous letting the secret slip can be when he ransacked stately Wayne Manor and broke Bruce's spine. With so many villains aware of Batman's public persona, the threat of retaliation toward those close to Bruce Wayne is ever-present. Plus, the large group of civilians who know Batman's identity makes him easy prey for those determined to learn who is underneath the mask and pointy ears. Even worse—it is hinted that Batman's greatest foe, the Joker, is aware of his civilian identity (but is probably too psychotic to care). Better beef up Wayne Manor security and add another padlock to the Batcave door.

THE DEVIL'S OVERDUE

COMIC BOOK HERO: Daredevil

SECRET IDENTITY: Matt Murdock, lawyer

ALLIES AWARE OF HIS SECRET IDENTITY: The Black Cat, the Black Panther, the Black Widow, Luke Cage, Captain America, Dr. Strange, Milla Donovan, Echo, Elektra, Falcon, Iron Fist, Jessica Jones, Lora, Moondragon, Nick Fury, Foggy Nelson, Ivan Petrovitch, Cecelia Reyes, Reed Richards, Sister Maggie, Spider-Man, Stick, Nathanial Taggart, Tork, and Ben Urich

ENEMIES AWARE OF HIS SECRET IDENTITY: Bull's-eye, Crossbones, Vanessa Fisk, John Garrett, Kingpin, Mephisto, Micah Synn, Mr. Fear, Silke, Typhoid Mary, and Voodoo Priest Mambo

TOTAL: 36

CHANCES OF BEING EXPOSED IN THE NEAR FUTURE: Inevitable. A tabloid once exposed Murdock as Daredevil, which led to him faking his own death. Recently, the villain Silke revealed that Murdock is Daredevil to cut a deal with the FBI. While the government military organization known as SHIELD was attempting to keep this information secret, a desperate agent sold out Daredevil to the *Daily Bugle* newspaper.

In response, Murdock publicly stated that he is not Daredevil. With the list of villains that know DD's identity and all the public outings, it's going to take a lot to keep the law offices of Murdock and Nelson open much longer.

* * *

Exposure can be a beast, but, as we'll find out in the next section, it pales in comparison to being faced with your kryptonite.

HOW KRYPTONITE WORKS

Just about everyone who grew up in the United States after the 1940s knows the core truth about kryptonite—it's bad news for Superman. Many also have an inkling that kryptonite comes in different colors. If you ask the right person, you'll find out that numerous varieties of kryptonite have existed over the nearly seventy years of Superman's history—which has included comic books, movies, radio dramas, comic strips, TV shows, and other media. Different iterations of the Superman franchise have used varieties of kryptonite differently (and unless otherwise noted, we're talking about the more recent depictions). Even if you restrict your question to the comic books, you can get wildly different answers. It all depends on the time period and title you ask about.

So what exactly is kryptonite? Where does it come from, and why is it here? What makes it so dangerous for Superman and other Kryptonians? And what does retcon have to do with it? In this section, we'll find out and learn what our own kryptonite might look like.

To understand how kryptonite works, it's helpful to know a few things about Superman:

- Superman's home planet, Krypton, orbited a red giant star called Rao, about fifty light-years from Earth. The planet was considerably larger than Earth, so it had a much greater gravitational pull.

- In comic books from the late 1930s, Kryptonians all had superpowers. However, in the current Superman universe, Kryptonians have no superpowers—Superman is only super because of Earth's weaker gravitational pull and its yellow sun.

Another critical thing to know about Superman is that a nuclear chain reaction in Krypton's core caused a massive explosion, destroying the planet. Through the years, explanations for precisely why this reaction occurred have differed:

- In some older comic-book story lines, Krypton had a uranium core.

- In *Superman #166*, a 2002 issue, Krypton was creeping toward its sun, and the sun's immense gravity pulled the planet apart.

- The modern explanation is that a great war took place on Krypton, and a doomsday device known as the Destroyer started the internal chain reaction that destroyed the planet.

⚡From Kal-El to Clark Kent: Superman's Rise and Radioactive Decay

Just before the explosion that destroyed Krypton, Kryptonians Jor-El and Lara outfitted their son Kal-El's birthing matrix for space travel. They sent Kal-El to Earth, where Kansas farmers Martha and Jonathan Kent found and adopted him. The Kents named the baby Clark, and he grew up to be Superman.

As Kal-El's birthing matrix traveled through space, it pulled fragments of the destroyed planet, made radioactive in the explosion, along in its wake. This radioactive debris became known as green kryptonite, or simply kryptonite, and it is deadly to superpowered Kryptonians. It does not react with oxygen, so it did not combust when it entered Earth's atmosphere. However, kryptonite is not indestructible—you can cut it, chip it, crush it, and melt it with acid.

When exposed to green kryptonite, superpowered Kryptonians instantly become weak. With prolonged exposure, they die. Green kryptonite has this effect because of the interaction between its radiation and the Kryptonians' cells.

On Earth, a variety of naturally occurring and manufactured substances emit radiation. They do this through one of three processes:

* **Alpha decay.** As an atom decays, its nucleus emits alpha particles, which are made of two protons and two neutrons.
* **Beta decay.** As an atom decays, a neutron in its nucleus spontaneously becomes a proton, an electron, and a subatomic particle called an antineutrino. The atom ejects the electron and the antineutrino, and the electron becomes a **beta particle**.
* **Spontaneous fission.** An atom spontaneously splits into two atoms of two different elements. It can eject neutrinos when this happens.

Atoms that undergo any of these processes often have lots of extra energy. They emit this energy as gamma rays, which are electromagnetic pulses—they're made of energy, not matter. Each of these forms of radioactive decay creates ionizing radiation, which can knock electrons off atoms. X-rays, another form of electromagnetic energy, are also a form of ionizing radiation.

Relatively speaking, alpha and beta particles cannot penetrate very far into matter. Gamma rays and X-rays, on the other hand, can penetrate matter, including human bodies.

Their ability to displace electrons from atoms can cause cells to mutate, sometimes causing cancerous tumors. Fortunately, lead blocks both gamma rays and X-rays. It's able to do so because of its high electron density. The rays are unable to penetrate the dense web of electrons found in a piece of lead.

Like radioactive Earth elements, kryptonite emits radiation, although exactly how kryptonite atoms decay is unknown. However, kryptonite radiation seems to behave like gamma or X-ray radiation—it can penetrate objects and living bodies but cannot penetrate lead. This suggests that kryptonite radiation is a form of electromagnetic energy, like gamma or X-rays, rather than particles of matter.

We'll look at how this radiation causes its deadly effects in the next section.

WHAT'S YOUR KRYPTONITE?

"Kryptonite" is one of those words that has absorbed a second meaning in common parlance. Some people use it to mean "something I absolutely cannot abide." HowStuffWorks staffers named the following as their kryptonite in response to an informal email poll:

✻ Tall ladders.
✻ Curry and bridges.

❋   Bell peppers.

❋   Small talk.

❋   Andie MacDowell and cilantro.

❋   Crappy comic-book movie adaptations.

❋   Email chains.

We all have different fears and dreads. What's your kryptonite?

⚡ Is Superman in Fact a Super-Plant? The Power of the Sun and Kryptonite

If Superman comes into contact with green kryptonite, he instantly becomes very weak. With enough exposure, he could die. Kryptonite has this effect because of the way it interacts with Superman's cells.

Much of Superman's power comes from Earth's yellow sun. His cells are like living photovoltaic, or solar, cells—they can store the energy from sunlight. Inside a photovoltaic cell, light comes into contact with a semiconductor, like silicon. The light's energy releases electrons from the silicon, and an electric field forces them to flow in one direction. In this way, a solar cell produces electricity.

You could also compare Superman to a plant that uses

photosynthesis to make its own food. Through photosynthesis, plants use energy from the sun to turn water and carbon dioxide into oxygen and sugar.

It's unclear exactly which method better describes the way Superman's cells use sunlight, if either of them describes it at all. It's also unclear exactly why light from a yellow sun affects Superman in a way that light from a red sun does not. If Rao were a red dwarf, the explanation could be simple—yellow stars are bigger, brighter, and hotter than red dwarfs. However, Rao is a red giant, meaning it is bigger, brighter, and a little cooler than Earth's sun. Although we do not know precisely why yellow light is so important, we do know that Superman needs it to have superpowers.

There are several possible explanations for how green kryptonite keeps Superman from getting power from the sun:

- The kryptonite radiation might displace the solar radiation responsible for Superman's powers.
- Kryptonite's ionizing radiation might displace electrons in Superman's cells, preventing the sort of electron movement found in solar cells.
- Kryptonite radiation may interrupt some other organic process within Superman's body.

As a result, Superman can no longer take advantage of the powers Earth's yellow sun gives him—he loses his powers, and he may die.

THE POWER'S SOURCE

The first Superman stories described Superman as a being from another planet and included no specific explanation for why he had super strength and other powers. Later stories explained that his abilities came from the difference in gravity between Earth and Krypton. Eventually, the explanation expanded to include Earth's yellow sun.

The History of Kryptonite

Green kryptonite is the most abundant and most frequently used form of kryptonite. But it hasn't always been part of the Superman universe, and it wasn't always green. An unpublished Superman story from the 1940s featured a precursor to kryptonite, a substance called K-metal.

The deadly element made its debut in the *Superman* radio series—*not* the comic book—in 1943. Its original purpose was to give voice actor Bud Collyer, who played the role of Superman, a vacation. With Superman incapacitated by kryptonite, another voice actor could supply his incomprehensible moans, filling the role until Collyer returned.

Kryptonite made another radio appearance in 1945 and appeared in a movie serial in 1948. But it didn't find a place in the comics until *Superman #61* in 1949, more than ten years after Superman's debut in *Action Comics #1*. Radio scripts and comic-book art portrayed the substance as red, gray, green, and metallic, but eventually the writers settled on green as the color of kryptonite.

In addition to "plain" green kryptonite, multiple varieties have appeared on the scene through the years. These varieties are different isotopes of the same element, and they can come in different grades, or strengths. The different isotopes have distinctly different effects on Superman and other life-forms, but these effects most likely all stem from the disruption of cells. Some varieties of kryptonite appeared in only one comic-book issue or story arc. For example:

- **Jewel kryptonite**, or Kryptonite 6, enhanced the powers of Kryptonians who had been sentenced to live in the Phantom Zone. It appeared in *Action Comics #310*.
- **X-kryptonite** appeared in *Action Comics #261*. It caused a cat named Streaky to develop superpowers.
- **Kryptonite plus** was an extra-potent variety that appeared in two story arcs.

YOU'LL SHOOT YOUR EYE OUT, SUPERMAN

Even though the substance known as kryptonite wasn't part of the Superman world until 1943, the word "kryptonite" had its debut years earlier. Daisy—maker of the Red Ryder air rifle—produced a Superman tie-in called the Krypto-Raygun in 1940. According to the ads, the rifle was made from "kryptonite, the amazing metal from his birthplace."

Types of Kryptonite

In addition, red, white, blue, and gold varieties of kryptonite were created through various means and appeared in the comic books until 1985. By the 1970s, kryptonite was everywhere—common criminals had pieces stashed away as protection from Superman. In *Superman #233*, an experiment gone wrong transformed all of the green kryptonite on Earth into iron. *Superman #255* eliminated it from the rest of the universe. But soon the radioactive mineral was back.

The year 1985 marked a turning point for Superman, kryptonite, and the entire DC Comics universe, which had become a multiverse full of alternate worlds. A miniseries called *Crisis on Infinite Earths* made major changes to the comic books' reality. The miniseries is an example of retroactive continuity, or retcon. Retcon is an attempt to clean up years of comic book history and get rid of unnecessary characters, plot holes, or in this case, whole universes. These changes are retroactive—the post-Crisis universe is not simply the way the DC Comics world works now; it's the way it has always worked.

In the post-Crisis reality, fewer versions of kryptonite exist, and the substance is relatively rare. In addition to *Crisis on Infinite Earths*, another recent retcon attempt was a miniseries called *Infinite Crisis*, but it did not substantially affect the use or presence of kryptonite.

So we now understand the substance that could destroy Superman. But is Superman invulnerable to everything and everyone besides kryptonite? Let's take a look at how old Supes would match up against some of our other favorite heroes and villains—including non-super ones like Chuck Norris and Keyser Söze!

BATTLING BLOCKBUSTERS: FIGHTING AGAINST SUPERMAN

I t began with an email from a HowStuffWorks. com reader named Ian Booth who asked, "Who would win in a fight—a Jedi Knight or Superman?" That was all it took for us to wonder how Superman would fare against other stars from movies, books, and comic books.

In this section, we'll look at what it takes to beat Superman and whether a range of foes could win against him. Like Superman, our opponents are all stars from big-name movies, books, and comic books. Let the games begin.

Superman vs. a Jedi

We'll begin with the question that started it all—would Superman win in a fight against a Jedi Knight? Booth's email raised several specific questions. Would a Jedi be fast enough to sense Superman's approach? Could the Jedi manipulate Superman in midair? Could a lightsaber be made from kryptonite?

Jedi are fast, and part of their training involves fighting

without using their sight or other senses. But they don't quite reach Superman's speeding-bullet status. Physically, Superman has the upper hand.

Because of Superman's strength and speed, using the Force to manipulate him mid-flight would be a lot like manipulating a starship. If the Force could manipulate starships in flight, Darth Vader could have crashed all the X-wings during the Battle of Yavin. Or Luke Skywalker could have destroyed all of the Star Destroyers during the Battle of Hoth.

The logical conclusion is that Jedi are not capable of stopping something as strong and fast as a speeding starship. Some portrayals, like the *Clone Wars* series, do show Jedi performing extreme feats with the Force, but those feats come with a price. To even attempt to stop Superman in mid-flight could seriously injure or kill a Jedi.

Finally, what about their weapons? Could Jedi use lightsabers made from kryptonite to destroy Superman? For better or worse, two things would make that highly unlikely:

* A lightsaber uses crystals to focus the blade, and gathering these crystals from the planet Ilum is a rite of passage for a Jedi. Under certain conditions, the crystals emit light and energy. Kryptonite is not typically crystalline; it's usually a metal or a rock. The crystalline

versions of kryptonite that have existed have emitted radiation, just like green kryptonite, rather than light or energy.

* Kryptonite affects Superman because of its radioactive properties. Adding kryptonite would only make a lightsaber more deadly to him if the resulting weapon emitted the same radiation, rather than being the source of the killing force.

However, a lightsaber might be useful against Superman even without the presence of kryptonite, since magical weapons can harm him. If the Ilum crystals are by nature magical, then the blade might slice through Superman just like it can slice through Vader's arm. With these points in mind, it seems unlikely that a Jedi could defeat Superman, regardless of who has the high ground, *unless* lightsabers are magical.

A Sith Lord, on the other hand, could be a formidable opponent for Superman. Unlike Jedi, Sith Lords don't hold themselves to a code of balance or fair play. They give in to powerful emotions like passion and rage, and they use the Force with a vengeance. Sith Lords also use synthetic crystals and make other adjustments to their lightsabers. They may be more likely to entertain the idea of making a kryptonite or magical lightsaber. Failing that, a Sith Lord might tune his lightsaber in such a way that it bypasses Superman's protective field.

A Sith Lord's no-holds-barred approach to wielding the Force also provides a potential advantage. So while he might not be able to stop Superman in midair, a Sith Lord could act first, unleashing the Force while Superman was on the ground. The ensuing fight between a Sith Lord and Superman has the potential to be long and brutal, especially as the Force weakens Superman. A Sith Lord could win, depending on:

- How effectively and frequently he can strike Superman with a lightsaber or Force lightning.
- Whether these weapons have an effect on Superman. If the Force and lightsabers can affect Superman, they could allow a Sith Lord to overcome the natural advantage Superman seems to gain when fighting evil.

POWER OF THE SUN AND PHONE BOOTH WARS

Superman's powers come largely from his exposure to Earth's yellow sun, and the sun has been charging his solar batteries for nearly all of his life. While it's unlikely at this point that Superman's batteries could run down, we'll assume, for the sake of simplicity, that all of these battles mentioned in this section take place on Earth or in the presence of a similar star.

Suppose Superman took on the other person best known for creative use of a phone booth—the Doctor of the BBC TV series *Doctor Who*. To survive, the Doctor would need to rely on his ship, the TARDIS (really shaped like a police box, not a phone booth). The TARDIS is a time machine, and with it, the Doctor could avoid the fight, prevent it from happening, or even prevent the planet Krypton from exploding.

⚡Superman vs. the Death Star

Let's look at the powers of these two first.

Superman has:

* X-ray and heat vision.
* Freezing breath and an impenetrable field that surrounds his body and his clothing.
* The ability to move at tremendous speeds and fly.

The Death Star has:

- 10,000 turbolaser batteries.
- 2,500 laser cannons and 2,500 ion cannons.
- 768 tractor beam projectors.
- A superlaser that focuses the energy from the station's hyper-matter reactor into a beam capable of destroying planets.

When fighting Superman, the Death Star's superlaser might be a liability instead of an asset. The reactor's power output is comparable to that of a small star. Depending on how similar this energy is to that of a yellow sun, the laser might actually make Superman stronger. But even without a power boost from the superlaser, Superman would still be invulnerable to the Death Star's other weaponry. Even if the Death Star managed to trap Superman in its tractor beams, it might not be able to penetrate his skin with its weapons.

In addition, Superman wouldn't need to find a two-meter wide thermal exhaust port in order to destroy the Death Star. His strength and dexterity would allow him to punch his way through the hull to access the space station's reactor. He could then use his heat vision to start a chain reaction before flying out again to escape from the resulting explosion.

To defend itself against Superman, the Death Star could leave the vicinity of any yellow sun and stay so far away that Superman couldn't reach it before losing his powers.

But this might only delay the inevitable—Superman's body has stored so much energy from Earth's sun that his batteries may never run out. Then, the Empire could retrofit its weaponry to use kryptonite ordnance. Otherwise, the Death Star would fall quickly.

We'll look at what would happen if Superman fought superheroes from a galaxy closer to home in the next section.

Superman vs. His Super Counterparts

The most logical people to pit against Superman are other superheroes. Even though they can't all beat Superman, many can at least hold him off for a while. Most of the superheroes we'd like to see go head-to-head against him are from Marvel Comics rather than DC Comics. The barrier between those worlds would have to fall for a while to let these brawls take place.

Some of the X-Men might be able to take on Superman without a lot of help. Here are a few examples:

* A fight between Wolverine and Superman, for example, would be like an unstoppable force meeting an immovable object. Superman is stronger and faster, but Wolverine has his adamantium skeleton and claws. Ultimately, the outcome depends on how strong and sharp adamantium really is and whether it can stand up

to Superman's strength and heat vision. Wolverine's healing factor might give him a slight edge, but not if Superman hurls him into the sun. The outcome of their fight would depend on luck, chance, and cunning.

* While Wolverine might hold his own while fighting Superman physically, Professor X and Jean Grey— even pre-Phoenix Jean Grey—are ideally suited to fighting him telepathically. Both are powerful psychics, and Superman has no defense against psychic attacks. Either character could defeat him by entering and destroying his mind, especially if they caught Superman by surprise.

* Another good opponent from the X-Men is Rogue, whose mutant abilities allow her to absorb ordinary people's strength and memories as well as mutants' superpowers. With a little luck, Rogue could touch Superman and absorb his powers. If she held on long enough, she could kill him, but it's unlikely that Superman would allow her to do so. However, the power Rogue absorbs may be the ability to gain strength from Earth's yellow sun rather than the physical manifestations of that strength. If that's the case, she would lose the ability she absorbed long before becoming strong enough to fight Superman. In addition, Superman is much faster than Rogue is, so she'd

need perfect timing or the assistance of other X-Men to distract him while she made her move.

With members like Rogue, Professor X, and Wolverine, it seems obvious that a team of X-Men could overpower Superman with ease. However, his speed, hearing, and physical invulnerability could make the battle difficult. The mission would require extensive planning and careful execution.

Another Marvel character who might be a worthy foe for Superman is Galactus, also known as the Devourer of Worlds. Superman can reverse the spinning of Earth to reverse time, but Galactus consumes entire planets. He is an immense being who can wield the Power Cosmic, a force so powerful that Galactus has to wear special armor to contain it. On a purely physical level, he's bigger, heavier, stronger, and more powerful than Superman. Beyond that, he's the Third Force of the Universe and keeps the balance between Eternity and Death. Fighting Galactus would be much like fighting a god.

However, if Superman is invulnerable to the Power Cosmic, it wouldn't matter how big or strong Galactus is. Superman would be able to outmaneuver him, and eventually Galactus's power would wane as he became hungry for another world to devour.

SUPERMAN VS. BATMAN, CHUCK NORRIS, AND KEYSER SÖZE

We could talk at length about whether Batman could defeat Superman. But lots of people have done so already, including Frank Miller in *Batman: The Dark Knight Returns*. A movie called *Superman vs. Batman* was even in the works as of 2014. It all comes down to the creative use of kryptonite, combined with Batman's cunning and technical savvy.

If Chuck Norris can turn up the sun when he gets cold, he can turn it down when he fights Superman, thus potentially turning Superman into a weak Clark Kent. Chuck Norris could wipe that guy out faster than you can say, "Walker, Texas Ranger."

Superman is big and strong, but Keyser Söze from *The Usual Suspects* is wily beyond description. Could he con the Man of Steel into doing something deadly? Maybe. But in spite of his skill at disguising his identity, Keyser Söze certainly couldn't convince anyone that he's really Superman.

Superman vs. the Lord of the Rings

In some ways, magic is the new kryptonite. Superman

has no defense against it, and it can affect him greatly. He's fought many powerful magicians, many of whom have been difficult to defeat. The outcome of these battles depends on strategy and exactly what kind of magical powers he's facing.

At first glance, Gandalf the Grey, a wizard from *The Lord of the Rings: The Fellowship of the Ring*, is no match for Superman. After all, he seems to be an elderly gentleman with a staff and a pointy hat who can make fireworks. But the important thing to remember is that Gandalf single-handedly defeated the Balrog known as Durin's Bane, an enormous, fiery demon inadvertently unearthed in the Mines of Moria.

Gandalf and the Balrog are both of the Maiar—powerful spirits present from the beginning of the world. The Maiar are second only to the Valar, the godlike beings who sang the world into existence. Gandalf isn't just an old man with a stick, and the Balrog isn't just a monster made of fire. They're both essentially demigods.

A QUICK LOTR RECAP

If you don't remember Gandalf's fight with the Balrog, here's how it went down. The foes met on the bridge of Khazad-dùm in the Mines of Moria. Gandalf stood between it and the rest of the Fellowship. He broke the bridge, and he and the Balrog both fell. Gandalf chased the Balrog for eight days, then fought it for two days and nights. Finally, he threw the Balrog from the top of a mountain, and its fall broke the mountainside. After the Balrog fell, Gandalf died.

The Balrog succeeded in killing Gandalf for one reason: when their battle began, Gandalf was already weary. If he faced Superman rested and refreshed, their battle would resemble his fight with the Balrog or Superman's first fight with Doomsday. Doomsday is the only being that has ever managed to kill Superman.

In fact, Gandalf and the Balrog are similar in much the same way that Superman and Doomsday are. Gandalf and the Balrog are both Maiar, and Superman and Doomsday are both superpowered beings from Krypton. Both pairs are like two sides of the same coin. Buildings crumbled around the brawling Superman and Doomsday, and a mountain peak turned into lava and rock as Gandalf and the Balrog fought.

A fight between Superman and Gandalf would be similar— the landscape around them would shatter from its impact. But, in the end, Gandalf is immortal, and he's nearly a god. Even though he doesn't have the youthful appearance or visually stunning powers that Superman does, he has the divine power to outlast and eventually overpower him. In addition, although he doesn't wield fireballs, Gandalf has power that most people would call magical and that can affect Superman.

⚡Superman vs. Harry Potter

Gandalf doesn't create many visual displays of magic, but Harry Potter does. Even though he's just a teenager and hasn't completed his magical training, Harry Potter has plenty of ammunition against Superman. Whether it's something as silly as the Bat-Bogey Hex or as devastating as Avada Kedavra, Superman has no defense against magic.

The real question is whether Harry would stoop to using one of the Unforgivable Curses on Superman. Here's how he could wield that wand in deadly ways against Superman:

* With the Imperius Curse, Harry could force Superman to do something deadly, like walk into a room full of kryptonite.
* He could also debilitate Superman with the Cruciatus Curse, which causes pain so intense that it can drive victims insane.
* Or, he could kill Superman directly with Avada Kedavra, the Killing Curse.

However, the three curses that could bring a swift and deadly end to a fight with Superman are also the three that Harry is least likely to use. All three are forbidden in the wizarding community. The only wizards who use them are dark wizards aligned with Voldemort. However, Harry has a Batman-like desire to avenge the killing of his parents. If Harry were convinced that Superman killed Lily and James Potter, he might react with the Unforgivable Curses.

Even without those curses, Harry knows plenty of spells that could slow down or stop Superman. A quick Petrificus Totalus would paralyze Superman completely. Various hexes and jinxes could also slow him down. But without an Unforgivable Curse, Harry wouldn't be able to end the fight. He'd need to enlist Hermione Granger, who could either find a source for kryptonite, a magical way to create it, or a spell to mimic its effects.

Harry would probably need both Hermione and Ron Weasley to help him strategize. Some spells require the caster to be able to see his target, and Superman's speed could make that impossible. Hermione, Ron, and Harry would have to work together to keep Superman from simply flying up to Harry and breaking his wand. For Harry to defeat Superman, he'd have to have the right motivation and very careful planning.

Regardless of the opponent, as we've seen in this section, any battle against Superman requires the right combination of luck, skill, strategy, and weaponry. Now that we have thorough examined this speeding superhero, let's check out another, far more human one with a lot of really cool technology and an odd phobia about a small nocturnal animal.

SUPERMAN VS. LOUD NOISES AND EVIL WILLOW

Superman's hypersensitive hearing isn't always an asset, since it makes him particularly vulnerable to sonic attacks. For example, he might find himself unable to defeat the Witch-king of Angmar, the leader of the Nazgul in the *Lord of the Rings* trilogy. The Witch-king cannot be defeated by man, which is why Eowyn, a woman, was able to destroy him. Superman technically isn't a man (he's a Kryptonian), but that loophole won't defend him against the Witch-king's piercing voice.

In season six of *Buffy the Vampire Slayer*, Willow Rosenberg gives in to dark magic. Earlier in the season, she demonstrated a flair for tinkering with time and dimensions. In the end, she flayed someone alive before trying to channel energy that would destroy the world. In addition to her immense power, Willow is unpredictable and grief-stricken, which would make her a formidable and reckless opponent. Most likely, Xander, often portrayed as the most useless member of the Scooby gang, would have to come to Superman's rescue in this case.

THE DARK KNIGHT IN SHINING ARMOR: HOW BATMAN AND HIS BATSUIT WORK

Superman may live by the sun, but night in and night out, Batman keeps the streets of Gotham City safe from crime and villainy. One of the most amazing things about Batman is this: despite the fact that Batman is a superhero, he doesn't actually possess any superpowers. In fact, Batman is extraordinarily ordinary. Instead of relying on unearthly abilities like Superman, he uses ingenuity, deception, and clever gadgetry to give him an edge over Gotham's criminal element. In the next two sections, we'll examine the exceptional science behind his two signature pieces: the Batsuit and the Batmobile.

Nothing instills fear in the hearts of Gotham's wrongdoers more than Batman's menacing appearance, and that appearance is all about the Batsuit. But the Batsuit does a lot more for Batman than make him look scary. On its own, the suit is an impressive piece of technology, combining armor, communications, and combat technologies into one state-of-the-art crime-fighting system.

But how exactly does the Batsuit do all that? Let's look at

its different components, how they work, and how Batman uses them to keep the streets safe at night.

⇟Suiting Up: The Armor

The main element of the Batsuit is a modified, advanced infantry armor system called the Nomax Survival Suit. Like many pieces of the Batsuit, the Nomax Suit was developed by the Applied Science Division of Wayne Enterprises. (The exact nature of Batman's relationship with Wayne Enterprises is unclear at this point.)

The foundation of the Nomax Survival Suit is a neoprene undersuit, much like a scuba diver wears. This undersuit basically functions as light, waterproof armor with temperature-regulating elements that maintain the body's temperature and keep muscles from freezing up in the field. There are six pieces of over-armor that attach to the undersuit:

- Knee guards.
- Calf guards.
- Leg armor.
- Arm guards.
- A full-torso vest.
- A spine guard.

Batman has augmented the over-armor with his custom-designed, scalloped-brass forearm gauntlets (painted matte black, of course). These gauntlets, along with his Kevlar gloves, make Batman's arms a viable defensive or offensive weapon. His gauntlets allow him to protect himself against bladed weapons like swords or knives. The armor and scalloped blades also add extra power and pain to Batman's strikes.

BATMAN BEGINS PRODUCTION NOTES

Throughout this section, we've included some cool facts about how the producers of the movie *Batman Begins* went about creating the real Batsuit and other awesome Batman technologies.

* "I looked at the great comics and graphic novels through the history of Batman to try and distill the essence of what those extraordinary pictures and drawings were saying about what Batman should look like."—Director Christopher Nolan on creating a new Batsuit.

* Because the matte black finish of the Batsuit got visibly dirty so easily, two crew members had to follow actor Christian Bale around the set, constantly wiping smudges off the suit during filming.

❋ The Batsuit for *Batman Begins* was created by costume designer Lindy Hemming (*Harry Potter and the Chamber of Secrets, Lara Croft: Tomb Raider*). Costume design was such a huge part of the production that the film crew built a village of trailers just for costumers and chemists to do their work in. This village came to be known as "Cape Town."

⚡ Helmets Are for Sissies: Batman's Notorious Hood

Batman replaced the traditional helmet that accompanies the Nomax Suit with his distinctive cowl, which does more than simply add to his daunting appearance.

Batman's cowl is sculpted from an impact-resistant, graphite-composite exterior. There is also bulletproof Kevlar plating that shields sections of Batman's head and protects it from small-arms fire. Bundled into the "ears" of the cowl are high-gain microphones and a radio antenna. The radio antenna allows Batman to monitor police and emergency-services radio frequencies so that he can always be "first on the scene."

The microphones in the ears are combined with special

earpieces in the cowl that give Batman superior hearing in the field. The microphones can also be used to amplify Batman's voice and broadcast it through a discreet speaker in the suit. This is what gives Batman's voice that distinctive, disembodied, unearthly sound.

BATMAN BEGINS PRODUCTION NOTES

* Unlike the Batsuits from previous films, the *Batman Begins* design allows the actor portraying Batman to turn his head. In the other films, Batman has to turn his entire torso to look to the left or right.

* During production, actor Christian Bale had one consistent assessment of the Batsuit: "It's hot, dark, and sweaty, and it gives me a headache." Costumers had this put on their crew T-shirts. Bale explains how he used that discomfort to his advantage: "It induced headaches and would send me into a foul mood after a half an hour...I used the pain as fuel for the character's anger...you become a beast in that suit."

⚡Swoooooosh: The Batsuit's Cape

Batman's cape was also developed by the Applied Science Division of Wayne Enterprises. Based on a nylon derivative, the special memory fiber can billow and flow like any fabric—but when an electric charge is applied to the fiber, it becomes stiff and takes on a distinctive shape. For instance, Batman's cape has been formed into a kind of glider than can slow his falls or even allow for short flights.

BATMAN BEGINS PRODUCTION NOTES

The real costuming challenge was creating a Batman cape with the right look and movement on screen. To accomplish this, the crew got help from the British Ministry of Defense. Ministry technicians taught the crew a technique called electrostatic flocking that is used on police gear to lower its night-vision detectability.

The crew brushed nylon parachute fabric with glue and dropped a very fine, hair-like material onto it. Then an electrostatic charge was run under the fabric. This process gave the cape a seamless look and a distinctive, velvety sheen. The idea was to make the cape in the film look more like what we've seen in the comic book.

⚡ Cool Contraptions: The Batsuit's Utility Belt

Batman's utility belt is a modified, Wayne Enterprises proto-type climbing harness. With an innovative gear-attachment system, Batman can grab and replace anything on the belt quickly and easily.

Batman carries a variety of nonlethal deterrents and other field equipment on his belt for crime fighting. The utility belt features:

* Magnetic grappling gun with monofilament decelerator climbing line.
* Flexible, fiber-optic periscope (to see around corners).
* Razor-sharp alloy, shuriken-style throwing stars (hand-ground into the distinctive Batarang shape).
* Ninja spikes for the hands and feet (used to climb sheer walls).
* Mini-mines and various nonlethal explosives (used to stun or confuse enemies).
* Mini cell phone with an encrypted signal.
* Medical kit.

Though it's not on his belt, one of Batman's most inno-vative and effective gadgets is the sonic device he carries in the heel of one of his boots. This device can be used to summon swarms of bats instantly to create mass chaos at any scene. This allows Batman to create hellish diversions or make dramatic escapes.

But as we'll see in the next section, Batman doesn't owe his dramatically successful escapes only to his Batsuit and its crime-fighting tools. In fact he wouldn't get very far without his infamous sidekick. We're not referring to Robin, though. We're talking about his megaton behemoth of a getaway vehicle: the Batmobile.

BATMAN BEGINS PRODUCTION NOTES

Before Christian Bale was cast as Batman, many other actors were considered or read for the role. These actors include Guy Pearce, David Boreanaz, John Cusack, David Duchovny, Hugh Dancy, Joshua Jackson, Eion Bailey, Billy Crudup, Cillian Murphy, Henry Cavill, and Jake Gyllenhaal.

TO THE BATMOBILE, ROBIN! HOW THE BATMOBILE WORKS

Whether you have seen the movie *Batman Begins* or not, you have probably seen the Batmobile. It is shaped like a spaceship with tires grafted onto it to make it street legal. The Batmobile used in *Batman Begins* is an icon for the movie and acts like a giant rolling advertisement for the film. But while the Batmobile exists on the page, is it an actual thing in the movie or just a computer-generated trick?

Perhaps surprisingly, the Batmobile is real. Every single time you see the Batmobile in the movie, you are seeing a real, physical object, not a computer-generated graphic. Whether it is driving on city streets at 100 miles per hour, landing in the Batcave, or pulling up to the scene of a crime, what you're looking at is a real car. When the Batmobile flies thirty feet through the waterfall to land in the Batcave, what's landing is a real, 5,000-pound vehicle. The Batmobile is so real that it actually served as the pace car for a major NASCAR race held in June 2005.

And yet, the Batmobile is also an illusion. Like so many

other Hollywood props, the Batmobile that you see in the movie does not exist at all.

How can that be? How can something be so real that it can serve as a pace car but also be so illusory that it doesn't actually exist? In this section, we will talk with Nathan Crowley—the man who designed the Batmobile and brought it to life in *Batman Begins*—to find out what's going on and how the Batmobile actually works.

⚡In the Beginning...

Let's start at the beginning and understand the "cinematic origins" of the Batmobile. In other words, let's understand how the car works in the movie.

* As we mentioned earlier, the first thing you have to understand about Batman is that he must have a car. Unlike Superman, Batman can't shout "Up, up, and away!" and fly through the air. Batman needs a serious set of wheels to get around—and get away when necessary.

* The second thing to understand is that, in this movie, Gotham City is portrayed as a highly dysfunctional version of New York City on steroids—there are surprises and obstacles at every turn. So Batman needs a *rugged* car.

* The third thing to understand is that Batman cannot, realistically, construct the car himself. Ordering all the

machine tools and parts and assembling them in the basement would give away his secret identity.

So in the script, they create a mothballed military vehicle built by Wayne Enterprises. Batman requisitions this vehicle for his own purposes and paints it black to match his color scheme. The Batmobile also gains some rather remarkable abilities. For example:

* It can go very, very fast.
* It has a jet engine that allows it to jump or fly through the air much farther than any normal car could.
* It has two driving positions—one for driving and one for jumping or flying.
* It has stealth capabilities, and part of the stealth mode is a silent, electric-motor drive.
* Getting into and out of the car is "unusual" to say the least. There are no doors—instead, the car "opens" somewhat like a flower.

Crowley is the man who had to take that cinematic vision of the car and bring it to life on film. To understand how he created this mega-machine, the first thing you have to understand about Crowley is that he is a very physical guy.

⚡ Getting Physical: Building the Batmobile

These days, the typical Hollywood way to handle a complex car like the Batmobile is to model it and simulate it with a computer. Even Yoda and Gollum are modeled and simulated on a computer, and a car is a piece of cake compared to Yoda.

This isn't how it works if you are Nathan Crowley. With *Batman Begins*, Crowley tended to be a staunch realist who wanted an actual, physical manifestation of the car in every frame of the film. Therefore, Crowley started the process of creating the Batmobile by model bashing.

Model bashing is a time-honored technique. You go down to a toy store, hobby shop, remote-control specialty shop, and the hardware store to buy parts—lots of parts of every size and shape imaginable. You buy lots of plastic models, toys, remote-control car kits, metal tubing, and so on.

You then cut and shape all of those parts to get the desired look for the car. For example, Crowley found that the nose cone from a plastic P-38 model kit made a perfect shape for the jet engine on the back of the Batmobile. So he cut off the nose cone, hollowed it out, added other parts to make it look like a jet, and glued it onto his model.

Crowley built six models like this, all 1:12 scale, before he got the look and the shapes that he wanted. This process took about four months.

Once he had the scale model, he started on a full-size replica.

⚡Sculpting the Batmobile

The next step was to build a full-size foam model of the car. So the Batmobile crew (including engineers Chris Culvert and Annie Smith, along with about thirty other people) took a gigantic block of Styrofoam and started carving it by hand.

They carved everything, including the rubber tires, in the foam. The goal of this process is twofold:

- First, to get all of the proportions right at full-scale. For example, this car is big—9 feet 4 inches (284 centimeters) wide. That's 8 inches (20 centimeters) wider than the typical eighteen-wheeler you see on the road. Getting the proportions right is important in something that big.

- Second, to have a full-size model that can be used to make things like the body-panel molds and the frame. The car has sixty-five separate body panels, and each one had to be manufactured on a custom-made wooden mold. The wooden molds were handmade from the foam model.

To make the steel frame, the Styrofoam model was cut up to get accurate sizing and panel mounting points for the frame. This sculpting process lasted about two months.

Now it was time to actually build and test the hardware. The crew built a "test frame" first…

⇆ The Batmobile Test Frame

As mentioned at the beginning of this section, the Batmobile that you see racing through the streets of Gotham City is a real car. To play its role in the film, this car had some amazing requirements:

* The car had to be able to go 100-plus miles per hour despite its size and heft.

* The car had to be able to accelerate from 0 to 60 mph in five seconds.

* The car had to be able to turn corners at speed. Lots of Hollywood cars can't turn or can't turn very well. To navigate the streets of Chicago (where the car scenes were filmed) at speed, this car needed to be agile. So it has a complete, real-world steering system.

* The car had to be able to jump up to 30 feet and then land completely unscathed.

To create this kind of performance, the team started with a steel "test frame" and put it through its paces. They worked on the engine, tuned the suspension, added special braking, and so on.

TRACTOR-STYLE TURNING

The braking in particular is interesting. To make the car turn, the team put extra brakes on each rear wheel and then mounted big hand levers on either side of the driver. To turn sharply to the left, the driver can brake the left rear wheel separately with the left-hand lever. This is much like the braking system used on tractors to help them maneuver sharply in the fields.

Once the team had the test frame running, it was time for the jump tests. The whole front end collapsed the first time around and had to be completely rebuilt.

Picking Out the Pieces: The Actual Components of the Batmobile

When they had the test frame performing the way they wanted, the basic configuration of the car and its drive train were set:

* The car uses a 5.7-liter Chevy V-8 engine. This engine has been tuned so that it can provide the power necessary to take a 5,000-pound vehicle from 0 to 60 mph (100 kph) in five seconds.

- The rear axle is a truck axle, with a truck transmission carrying power from the engine to the axle. The truck axle added a lot of weight to the vehicle. They wanted the car to be as light as possible so it would jump better, and this axle was the opposite of "lightweight." That extra weight is one of the things that contributed to the strain on the front end in the first jump tests.

- The rear tires are 37 inches in diameter. They are off-the-shelf, 4x4 mud tires called Super Swampers, made by Interco.

- The front tires are racing tires made by Hoosier.

- The front wheels have independent suspension elements inspired by the long-travel suspensions of Baja racing trucks. When airborne, the front wheels pop out about 30 inches on their suspensions to absorb the shock of a 30-foot fall.

By this time, the design and development process had taken about nine months and consumed several million dollars. However, the payoff was high, because now the team could begin manufacturing Batmobiles on an assembly line.

⚡ Building the Batmobile: The Assembly Line

Besides the test frame, the team manufactured four complete, street-ready race cars. To do that, they built the steel frames and mounted the drive trains on each one. Then the body shop manufactured the sixty-five carbon-fiber panels for each car.

NASCAR INTERIOR

The "race" versions of the Batmobile are the cars that careened through the streets of Chicago during filming. From the outside, they look like Batmobiles. Inside, however, they look like NASCAR race cars.

According to Crowley, when you get in the car, what you see is the steel frame of the car along with sheet metal covering some of the surfaces, as in a NASCAR car. The gauges are all exposed. There is a Halon fire-suppression system along with other safety features to protect the drivers.

Visibility in these monsters was terrible. The driver could see out the front window fine, but there was no side or rear visibility. So the team mounted side and rear video cameras, and the driver used monitors to see outside. Because of these extenuating circumstances, the drivers for these Batmobiles trained for six months before they started driving on the streets of Chicago.

So why'd they build four complete Batmobile race cars?

There were two reasons. First, the team expected there to be accidents and wanted to have multiple cars in case one or two wrecked. Think about it: these cars have the ability to go 100 miles per hour but have hand levers to help turn corners. They also are called on to jump 30 feet.

Accidents seemed likely. The good news is that no accidents actually occurred (if you ignore the incident in which a driver rear-ended one of the Batmobiles as the four cars were moving to a new location). The six months of training and the drivers' skill really paid off.

The other reason is that two of the four cars are special:

- One is the **flap version**. It has all of the hydraulics and flaps to handle the close-up shots where the car is "flying."
- The other is the **jet version**. Crowley didn't want to "add on" the jet flame with a computer—he wanted a real jet flame. The car has an actual jet engine fueled by six propane tanks located inside the car. The team can mount and dismount the jet as needed for filming.

MEGA MONEY FOR MEGA MACHINES

Each of these four cars cost about $250,000 to build. You could buy four Bentleys for the cars' combined cost! But that pales in comparison to the millions of

> dollars spent from start to finish to research, develop, and fabricate all of the models that made the Batmobile possible.

These four cars worked great for the street scenes. However, when you watch the movie and Batman gets in and out of the car, you obviously do not see him getting into a vehicle that has a bare frame with a Halon safety system and riveted sheet metal. The interior of the Batmobile is cool. Two other teams made the Batmobile's interior possible.

⚡Step Inside the Batmobile

When you watch the movie one of the most interesting parts of the Batmobile is the way that Batman gets in and out. It is almost like a flower opening—the roof unhinges, the windshield slides back, and the seats in the car actually rise up. To make all of those origami-like folds, a separate team built yet another Batmobile.

This car is the one that Batman "pulls up in." It has several unique features:

* It is loaded with hydraulics to make the opening and closing of the cockpit happen in a realistic way.

* It has a small electric motor that lets the car drive forward, but there is no massive V-8 engine and no need for the car to drive at a high speed.

* The car actually has another driver hidden inside the vehicle. He makes the car stop and start as needed for each shot.

But when you see Batman inside the car, yet another piece of the puzzle appears. The interior of that car is actually a studio set that can't move at all. It is oversized so that cameras can fit inside, and it has all of the features needed to shoot the "interior shots"—things like the seat that can move forward, the cockpit controls, and so on.

And finally, there is one other version of the Batmobile—the miniature version. It is a 6-foot-long (2 meter), 1:5 scale model of the Batmobile, complete with an electric motor drive. When you see the Batmobile flying through the air across ravines or between buildings, it is this scale model that does the flying. (But it's the 5,000-pound, race-car version that flies through the waterfall to land in the Batcave.)

So now you can start to see the complete illusion that makes the amazing car known as "the Batmobile" possible in the film.

⇄Understanding the Illusion

So let's put all of the pieces together to understand the illusion:

* When you see the Batmobile careening through city streets, that is one of the four "race" versions of the Batmobile. They have real engines and drive trains, but the interior is a stripped-out permutation of a NASCAR race car.

* When you see the Batmobile fire its jet engine, you are seeing the special "jet version" of the car. There are six extra propane tanks hidden inside the car to fuel the jet.

* When you see the Batmobile flying through the air, that is usually the 1:5 scale miniature version, intercut with shots of the "flap version" of the full-sized car.

* When you see Batman get in and out of the car, that is the "opening" version of the Batmobile. It has a more realistic interior and a separate driver hidden inside the vehicle.

* When you see Batman inside the cockpit, that is a scene shot in a static set.

Here's an example of what happens in the film when Batman drives home to the Batcave after a long day of crime fighting:

1. The race version drives down the road.

2. The jet version fires its jet engine.

3. The miniature version flies across the ravine.

4. One of the full-sized race versions enters the Batcave, flying through an actual, full-size waterfall. This shot is incredibly complex. They turn on the water for the waterfall. The car drives down a road and shoots off the end of a ramp to get itself airborne. The car travels through the waterfall, where the downdraft and weight of the falling water has to be taken into account to get the angles right. The 5,000-pound car flies 30 feet and lands with an incredible thud on a reinforced-concrete landing pad. A sand berm helps slow the car down. There is also a huge arresting cable to stop the car in case something goes wrong.

5. The view switches to the static cockpit set to show Batman's perspective.

6. The view switches to the "opening version" of the car, which pulls up in the Batcave set, stops, and unfolds so Batman can get out.

All of those different, real, physical versions of the car come together in the movie to create the illusion of the Batmobile.

As Crowley points out, understanding this process gives you an appreciation for why a modern film can cost so much to make. The Batmobile itself is eight different versions of the same car, built and managed by several teams containing dozens of people. The "Batmobile" cost many millions of dollars in research, development, and fabrication. Long story short, we're not ready to manufacture Batmobiles for everyday use. But that doesn't mean you won't see them careening along the streets one day, hopefully on real-life crime-stopping missions.

QUIZ: THE ULTIMATE BATMOBILE QUIZ

The Batmobile is a lean, mean driving machine. But it takes serious skills to maneuver (and fly!) this killer car, let alone build it. Think you can do it? Let's test your knowledge to see whether you would succeed as a Batmobile handler with The Ultimate Batmobile Quiz. Check your answers in the back of the book on page 275.

1. **How much does the Batmobile weigh?**
    a. 800 pounds
    b. 2,000 pounds
    c. 5,000 pounds

2. **Which of the following is not a Batmobile driving position?**
    a. Driving
    b. Jumping/flying
    c. Swimming

3. **What is model bashing?**
    a. The process of creating a model for a

vehicle using parts from toy cars, R/C cars, and plastic models.

b. A process to make a car using scraps from other automobiles.

c. A process by which a model car is computer generated and then built to scale.

4. **The Batmobile's creator built the model for the real Batmobile on a _____ scale.**

a. 1:12

b. 1:50

c. 1:100

5. **After a scale model was designed, a full-size model of the Batmobile was built from:**

a. Aluminum

b. Styrofoam

c. Clay

6. **To play its role in the film *Batman Begins*, the actual Batmobile had to accelerate from 0 to 60 miles per hour in:**

a. Three seconds

b. Five seconds

c. Eight seconds

7. **In what city were the car scenes involving the Batmobile filmed for *Batman Begins*?**
   a. New York
   b. Los Angeles
   c. Chicago

8. **The Batmobile's rear tires measured:**
   a. 37 inches
   b. 30 inches
   c. 24 inches

9. **How much did each street-ready Batmobile used in filming cost to build?**
   a. $100,000
   b. $250,000
   c. $520,000

10. **Which of the following models was used in the scenes in which the Batmobile travels between buildings?**
    a. A computer-generated Batmobile
    b. The 5,000-pound, full-size racing model
    c. The miniature version, a 1:5 scale model that uses an electric motor drive

NO LAUGHING MATTER: HOW THE JOKER WORKS

"Once again a master criminal stalks the city streets—a criminal weaving a web of death about him—leaving stricken victims behind wearing a ghastly clown's grin—the sign of death from THE JOKER!"

—*Batman #1*, 1940

G reen hair, white skin, a red grin, and an unlimited supply of purple suits—when a certain playing card made its appearance at the end of the big-screen relaunch of *Batman Begins*, fans started clamoring for every grainy image or leaked plot detail about everyone's favorite Bat-villain. The Joker, played by Heath Ledger, made his return to the big screen in *The Dark Knight* in the summer of 2008. But long before the Joker was mixing it up in big-budget summer movies, he was making Batman's life miserable in the monthly titles of DC Comics' *Batman* and *Detective Comics* series. Who exactly is this master of murderous mischief, and how does he get away with what he does?

⚡A Dangerous Debut: The Joker's First Appearance

The Joker was originally conceived during the Golden Age of comics as an evil court jester. He made his first appearance in *Batman #1* (1940). The original Joker resembled actor Conrad Veidt as he looked in the silent film *The Man Who Laughs*.

Baffling both Batman and the Gotham Police Department, the Joker made his debut as a master thief and mass murderer. He announced the name and exact time of death of his next victim over the radio. The Joker accomplished this feat by poisoning his victims with a time-released, facial-contracting poison known as Joker Toxin.

In this first portrayal, the Joker was a calculating killer who left victims with a permanent death grin. While he was slated to be killed off from a knife wound in his second appearance in the same issue, the DC editorial department felt the character had potential. At the last minute, a panel was added at the end of the comic revealing that the Joker was still alive. This started the trend of having the Joker meet his apparent demise only to be revealed as alive and well.

A JOKER BY ANY OTHER NAME

Throughout his comic book career, the Joker has been referred to as the Crown Prince of Crime, the Harlequin of Hate, and the Ace of Knaves. As for his real name, the Bat-story *The Return of Hush* reveals it to be Jack, though the legitimacy of this is questionable. His last name is unknown.

⚡ The Joker's Comic Origin

"I'm not exactly sure what happened. Sometimes I remember it one way, sometimes another...If I'm going to have a past, I prefer it to be multiple choice!"
—The Joker, *The Killing Joke*, 1988

In comics lore, as revealed in *Detective Comics #168* (1951), the man who would become the Joker first masqueraded as a criminal known as the Red Hood. During a botched robbery at a Gotham chemical plant, the Hood jumped into a vat of hazardous chemicals to escape from Batman. Upon emerging from a drainage pipe, he discovered that his skin had been bleached white, his hair had turned green, and his mouth was permanently distorted into a large grin with bright red lips. The chemicals caused him to go insane, thus transforming him into the Joker.

Although this origin is considered canon in DC Comics

continuity, the Joker's earlier life is up for debate. So are the circumstances surrounding why he wore the Red Hood guise. The most accepted origin is found in Alan Moore's *Batman: The Killing Joke*. Moore presents the pre-toxed Joker as a chemical engineer who quits his job to become a stand-up comedian. Failing miserably as a comic and desperate to provide for his pregnant wife, the man accepts the job of guiding two thieves to the Monarch Playing Card Factory through the neighboring chemical plant where he worked.

During the planning stages of the heist, police inform the man that his wife has been killed in a household accident. It's possible she was killed by corrupt police involved in the heist. Although he's hesitant to continue the caper, the man is strong-armed into donning a red helmet and leading the two thugs through the plant.

While *The Killing Joke* is the most accepted comic-book origin of the Joker, the desperate engineer is not the only version of the villain's beginnings. In another retelling, the young Joker collects the bones of animals he tortures. He also takes pleasure in his mother's abuse at the hands of his father. This version reveals that the Joker's first murder was committed on a neighborhood boy who discovered his "pet cemetery" and that the young Joker later murdered his father.

In addition to the different stories of how the Joker came to be, there is much debate over who should be called the father of the Clown Prince of Crime. Comic-book artist Bob Kane, fellow artist Jerry Robinson, and writer Bill Finger all stake a claim to having created the Joker. Kane maintains that he and Finger created the Joker, while Robinson contributed only the Joker's playing card. Robinson, on the other hand, alleges that he had already drawn the Joker before Finger ever showed him a picture of Veidt.

Next, we'll look at changes to the Joker's character over the years.

⚡ Calming the Clown

"Batman! I accuse you of interfering with the right of criminals to commit crimes!"

—*Batman #163, The Joker Jury*

During the 1940s and 1950s, comic books featuring horror and criminal themes were among the most popular titles being published. Believing that these books were helping to fan the flames of juvenile delinquency in young men, Frederick Wertham published *Seduction of the Innocent: The Influence of Comic Books on Today's Youth* in 1954. In his book, Wertham reprinted gruesome, violent, and sexually thematic images from popular comics. He argued that comics encouraged antisocial and homoerotic behavior in American youth.

The public responded with disdain toward the comics industry. Many forbade their children from reading the books. To stay in business and win back the trust of the general public, the major comics publishers created the Comics Code Authority and began self-regulating the content of their books. Comic-book publishers had to abandon the zombie-, monster-, and crime-themed books that were their chief moneymakers in favor of less popular comedic and toned-down comics. Many publishers had to close their doors. If not for the successful resurrection of superhero comics by Marvel and DC in the 1960s, the comic book might have vanished altogether.

As a result of pressure from comic-fearing parents and the creation of the Comics Code Authority, the Joker was portrayed as a more comedic figure during the 1940s and 1950s. His tone was shifted from that of a crafty, dangerous murderer to a wacky, annoying prankster. He became a mischievous thief who concocted elaborate heists that often involved intricate puzzles and disguises.

While the Joker now practiced a more nonlethal array of activities, he remained one of Batman's most intelligent adversaries. He employed numerous gimmicks such as "Crime Costumes," Joker utility belts, and even a Joker-Mobile that featured a large Joker face on the front grill to help further his criminal schemes. This version of the villain lasted until the 1960s. Then, Julius Schwartz, who was not a fan of the Joker character, started editing the Batman comics. The Joker nearly disappeared into comic-book oblivion.

⇒ Modern Mania: The Joker Today

"Zip-a-dee-do-da, zip-a-de-oom, my-o-my, what a wonderful boom..."

—*Joker #3, The Last Ha Ha*

When writers Denny O'Neil and Neal Adams were tasked with updating the Batman line in 1973, the Joker came back to the forefront. In his latest incarnation, the Joker returned to his more murderous ways. But instead of a cold and calculating thief, this Joker became a homicidal maniac with no reservations about impulsive and sporadic killings.

This more lethal Joker envisioned himself as Batman's equal in terms of intelligence and wit but did not hesitate to employ the occasional gimmick, like Joker Toxin. He displayed increasingly sociopathic behavior, such as continuously murdering his henchmen. It was during this time that the Joker even received his own limited series. In it, he battled a different hero each issue only to be re-incarcerated at the conclusion of each story.

During the late 1980s, this maniacal behavior was taken

to the next level in several episodes. His crimes included beating the second Robin, Jason Todd, with a crowbar before apparently killing him in an explosion. He also kidnapped and tortured Commissioner Gordon and paralyzed Batgirl, Gordon's daughter Barbara, by shooting her in the spine.

In current DC continuity, the homicidal harlequin incarnation of the character has remained firmly in place while the Joker has engaged in some of his most ambitious endeavors to date. Such campaigns include stealing Superman foe Mr. Mxyzptlk's reality-altering powers to reshape the world in his image, as well as infecting a large population of supervillains with his insanity-producing Joker Venom.

The Joker has also managed to find a sidekick of sorts in Harley Quinn, his one-time psychiatrist, who he manipulated into becoming obsessed with him. Though occasionally thinking on a grand scale, the Joker has never become a stranger to random sadism. This is evident in the murder of Commissioner Gordon's second wife and the shooting of Zatanna, magician and Batman ally.

Recently, the Joker was shot in the face by a corrupt police officer masquerading as Batman. He nearly died before being resuscitated at Arkham Asylum. Extensive reconstructive surgery was required to revive him, and now the Joker has been left with a permanent grin that makes it difficult for him to speak.

Believing himself to have been shot by the actual Batman and once again facing further disfigurement, the Joker emerges with an even more lethal and psychotic persona. Able to communicate by blinking his eyes in Morse code, the Joker orders Harley Quinn, masquerading as his speech therapist, to execute each of his former henchmen. Sparing only Harley, the newly dubbed "Crown Prince of Killers" and "Harlequin of Hell" now believes that he must up the ante to compete with Batman, who he believes is willing to murder him.

⚡ Last Laughs

"I did it! I finally killed Batman! In front of a bunch of vulnerable, disabled kids!!!! Now get me Santa Claus!"
—*Batman #655, Batman and Son, Part 1*

The ongoing battles between Batman and the Joker raise a couple of questions. Why haven't the Joker and Batman killed one another? And if the Joker has murdered so many people, why is he still alive?

As far as Batman is concerned, it is against his personal code and mission statement to kill his enemies and thus become as evil as they are. The Joker's excuses, on the other hand, are a little more complex. On numerous occasions, the Joker has gotten the drop on Batman only to refuse to finish his foe. His reasons include it not being a dramatic enough ending to their rivalry or feeling that putting a prone Batman down would be anticlimactic.

The Joker feels that it is his destiny to defeat Batman in a worthy manner that finally proves his random madness is superior to the Batman's calculating detective skills. It has also been hypothesized that the Joker fears that without Batman as a catalyst for his actions, his life would become meaningless, as evidenced in *Batman #663* when the Joker states, "I could never kill you. Where would the act be without my straight man?"

The Joker has managed to avoid the death penalty due to being labeled criminally insane. In the graphic novel *Arkham Asylum*, writer Grant Morrison proposed that the Joker suffered from a kind of "super-sanity" of heightened sensory perception and that he lacked a true personality of his own, adapting his psyche to whichever was the most beneficial. Morrison revisited his ideas on the Joker's sanity in *Batman #663* and proposed that each time the Joker escaped from confinement, a new personality would emerge, explaining the changes in the his character through the years.

Thus, instead of being sent to prison following his capture, the Joker is sent to the Gotham City psychiatric hospital, Arkham Asylum. Originally created to house only

Batman's mentally disturbed foes such as Joker and Two-Face, Arkham Asylum has evolved into the destination for most of the Dark Knight's rogues' gallery. It's also where the Joker met his ex-shrink turned sidekick, Harley Quinn. And as we'll see in the next section, it's possible that Batman might also belong there in light of certain behaviors of his.

THE MAN BEHIND THE MASK: IS BATMAN A SOCIOPATH?

We've taken a look at the two main supertechnologies that make Batman go; now let's look at the man behind the mask. For years, citizens of Gotham City have relied on the masked avenger known as Batman to protect them from the schemes of numerous madmen, including the Joker, the Scarecrow, and Mister Freeze. But Batman's efforts to defend the city haven't been without controversy. Critics have accused him of everything from endangering the populace to behaving like a madman. Is it possible that Gotham's greatest champion is a sociopath, like so many of the inmates he's sent to Arkham Asylum?

WHAT IS A PSYCHOPATH OR SOCIOPATH?

People who are considered to be psychopaths or sociopaths generally suffer from antisocial personality disorder. This is a psychiatric condition characterized by chronic behavior that manipulates, exploits, or violates the rights of others. This behavior is often criminal.

To properly speculate about whether Batman suffers from antisocial personality disorder, you must first look at his basic behavior.

* Batman costumes himself and engages in vigilante activities to control the Gotham criminal element.
* He often uses physical force and weapons to subdue those he deems lawbreakers, but he doesn't fatally injure the people he apprehends.
* Batman is extremely secretive and generally operates alone, though he's rumored to have worked alongside the Gotham City police department.
* Batman shields his features to avoid identification and is most likely a resident of Gotham City himself.

That said, do Batman's activities make him a candidate for antisocial personality disorder (APD)? Let's take a look at the criteria from the American Psychiatric Association and see how Batman relates. (Note: He must meet three criteria in order to be diagnosed with APD.)

* **Failure to conform to social norms with respect to lawful behaviors as indicated by repeatedly performing acts that are grounds for arrest: CHECK.** While Batman often works alongside the Gotham City

police department and is rumored to have direct communication with Police Commissioner James Gordon, his vigilante acts are technically illegal. While there is no warrant out for his arrest, Batman is not a licensed law-enforcement officer of Gotham City and routinely ignores laws pertaining to vigilante justice.

* **Deceitfulness as indicated by repeated lying, use of aliases, or conning others for personal profit or pleasure: NO.** Batman simply uses his disguise to conceal his identity, not to benefit maliciously at the expense of others.

* **Impulsivity or failure to plan ahead: NO.** Batman displays the opposite tendency: he is a master tactician and doesn't appear to act on impulse, but rather employs careful strategic planning to confront an issue or situation wherever possible.

* **Irritability and aggressiveness, as indicated by repeated physical fights or assaults: CHECK.** Batman is reported to have assaulted scores of individuals in his pursuit of information related to criminal activity. He shows heightened aggression with those he deems criminal, using both his body and external weapons to subdue or incapacitate his targets.

* **Reckless disregard for the safety of self or others: NO.** Batman has ceased pursuit of a target numerous

times to prevent endangering the lives of others. Also, while he constantly places himself in dangerous situations, he never does so in a reckless manner.

* **Consistent irresponsibility, as indicated by repeated failure to sustain consistent work behavior or honor financial obligations: NO.** Batman's efforts to thwart the criminal underworld have been consistently strong. Also, while it's not clear if he has steady employment in his civilian life, he certainly does have a steady stream of income.

* **Lack of remorse, as indicated by being indifferent or rationalizing having hurt, mistreated, or stolen from another: NO.** Batman has no record of theft. While he will use force and cause pain when apprehending criminals, he is not indifferent to their future well-being, as he doesn't use more excessive force than necessary.

* **Insensitivity to pain: NO.** Batman is extremely sensitive to the pain of the citizens of Gotham City and constantly works to ease the oppression of upright citizens by the underworld. He appears to be driven to help those who cannot help themselves.

By applying these guidelines to the information that's readily available on Batman, it appears that he exhibits only two symptoms of the disorder. In other words, he doesn't meet the criteria to be called a sociopath. Of course, without the knowledge of his civilian identity and the cooperation of Batman himself, it's impossible to validate the accuracy of this profile. Regardless of whether the Batman persona is a sociopath, his secret identity may be.

MAY THE FORCE BE WITH YOU: HOW DOES A *STAR WARS* LIGHTSABER WORK?

Now that we've looked at some superheroes and supervillains on Earth, let's take to the skies! How do superpowers work up there? In the next few sections, we'll take a look at the science behind waging intergalactic wars and trekking among the stars. Let's start with George Lucas's sci-fi canon *Star Wars* and his world-famous weapon, the lightsaber. How do they work in real life?

Like the Millennium Falcon and Yoda (and even Chitty Chitty Bang Bang, for that matter), lightsabers are a special effect that looks so real you actually believe they exist! Unfortunately, lightsabers are not real. In our world, there is no such thing as the diatium core that produces the energy to power lightsabers, as in the films. So if there's no such thing as diatium, let's take a look at the technology Lucas and others had to use to produce these illusions.

The technique used to create the lightsaber effect is fairly straightforward but disappointingly tedious.

- On the set, the actors use props composed of handles

that have aluminum rods attached to them, and these rods are the length of the lightsaber "blade."

* The handles are plastic models, and the aluminum rods are painted red or green or blue. The actors use these props as though they were lightsabers.

After the film is shot, it is taken to the special effects department. The film is developed normally. In this film, the actors look like they are fighting with painted broomsticks instead of lightsabers. What's to be done now to get that cool laser effect?

* A special-effects artist has the job of making those broomsticks look real. The artist looks at the film frame by frame, and projects each frame that contains a lightsaber onto a clear piece of plastic (an animation cel).

* The special-effects artist then draws the outline of each lightsaber blade in the frame onto the cel.

* Then, for each frame, the artist paints in the correct color for the blade using a bright cartoon color.

* Eventually the artist has a stack of these cels, one for each frame of the movie containing a lightsaber. The cels are clear everywhere except where the lightsaber blade is seen in each frame.

Now, a new piece of movie film is shot. On this film, each animation cel is placed over a black background and shot with a light diffuser over the lens (this diffuser gives the lightsabers the glow they have around the edges).

If you were to play this film in a projector, all you would see is the lightsaber blades moving on a black background.

Before it is developed, however, the actual footage from the movie is double-exposed onto this same film. The effect is amazing—the lightsabers look incredibly bright and realistic!

As movies move further into the digital realm, the job of animating the lightsabers gets slightly easier, but not much. In a digital world, each frame of the movie is scanned into a computer at extremely high resolution so that each frame can be manipulated on a computer screen. To make the lightsabers look real, the special effects artist looks at each frame on the computer screen, outlines the broomsticks, colors the areas, and diffuses them (frame by frame by frame…). Instead of being done on a plastic cel, it is all done on separate "cels" in the computer's memory and then merged digitally. However, there is no way to get around the fact that the animator must look at each frame and outline the lightsaber blades one by one.

So now that we've looked at the somewhat tedious real explanation of lightsabers, let's have some fun thinking up how we could use these cool tools beyond fighting off those pesky Sith Lords!

DARTH VADER'S DRIVING FORCE: THE SITH EXPLAINED

Have you ever wondered exactly who the Sith are, where they came from, and why they are considered so evil? Why is it that, as Yoda says, "Always two there are, no more, no less"? Why are their lightsabers always red, and what's with all this vengeance? What are they trying to avenge anyway?

It turns out that, as with all things Lucas, there is a fairly huge story behind it all that has played out in movies comics, novels, video games, and various other media over the past thirty years. Once you really sift through all there is to know about the Sith, you'll find there's quite a bit to these guys. What you see in the movies is far from the first or worst time the Sith and the Jedi have squared off. Let's get started.

⚡ "Always Two There Are—a Master and an Apprentice": Sith Basics

The Sith are an order of Force practitioners who have chosen the dark side of the Force as the source of their power. Members of the Sith Order are called Sith Lords or

Dark Lords of the Sith. The Sith are the sworn enemy of the Jedi, whom they have fought for thousands of years.

Sith only operate in pairs—one master and one apprentice. Though it was not always so, under the current doctrine of the Order there can only be two Sith Lords at any given time. This doctrine was put in place by a legendary Sith Lord named Darth Bane. Bane did this to prevent the destructive infighting that had plagued the Sith Order for millennia. With this system, Dark Lords pass the Sith legacy through the generations in an unusual and violent way.

After years of working together, an apprentice will kill his master when he senses an opportunity to become the master himself. This new master then takes an apprentice of his own. Of course, should the apprentice fail to kill his master, he will surely be killed in retaliation and the master will find a new apprentice. This method uses the Sith Lords' innate ambition to make sure that only the strongest Sith survive.

Unlike the Jedi, who use their own names, Sith Lords take a title that starts with Darth (the ancient Aurabesh pronunciation for the word for "dark") and that ends with a name of their choice, such as Darth Vader or Darth Stuffo. This was not always the case with the Sith. It's a system used by modern Sith Lords to hide their true identities.

Like the Jedi, the Sith are skilled in the art of lightsaber combat. Some of the most legendary Force swordsman have been Sith Lords. In recent history, Darth Maul was considered one of the most talented and savage warriors ever to wield a lightsaber. Maul used a custom-built, double-bladed lightsaber and could easily fight several Jedi at once.

In the Jedi Order, each Jedi student makes a journey to get Ilum crystals for the lightsaber he will one day build. This journey is considered a rite of passage. As a result, Jedi sabers are many different colors, depending on the crystals the Jedi find. There is no such ritual for the Sith. The Sith prefer synthetic crystals for their sabers. The synthetic crystals save the Sith the long trek to get Ilum crystals, provide more blade-tuning options, and give Sith sabers their characteristic red blade.

After thousands of years and many wars, the modern Sith Lords have learned to use deception and subterfuge rather than outright warfare to further their quest. This shadowy, manipulative approach is what ultimately led the Sith Order to destroy the Jedi Order and take over the galaxy during the rise of the Galactic Empire. But as with all things, none of this would have been possible without the boundless power afforded to them by the Force.

DUAL WIELDING

Darth Maul was not the first Sith Lord to wield a double-bladed lightsaber. Centuries before, Exar Kun, a Dark

Lord of the Sith, was the first person known to wield this insanely dangerous weapon. The use of a double-bladed saber is a deadly and esoteric saber-fighting style. No other Jedi or Sith has ever been known to use a double-bladed lightsaber on a regular basis, though the Jedi use them with training droids to simulate multiple attackers.

While he did not use a double-bladed lightsaber, the soon-to-be Sith-Jedi student Anakin Skywalker used two lightsabers (one in each hand) in his first duel with Darth Tyranus. He eventually lost the second lightsaber in the duel, and his arm along with it.

⚡ Use the Force

The Force is the power in the galaxy that binds everything together. Its energy flows through the blood of all living things. Certain individuals and creatures have a natural adeptness or Force sensitivity. When Force-sensitive individuals tap into this, they can unlock unimaginable power within. A powerful Force practitioner can move objects, predict the future, bend wills, move with lightning speed, and practically manipulate reality itself through the use of the Force.

The Force is divided into two sides: the dark side and

the light side. Each side of the Force has its unique powers, strengths, and weaknesses. Likewise, Force practitioners are divided. The Jedi use the light side of the Force, while the Sith use the dark side. Ideologically speaking, two questions draw the line that divides the Force:

* What is the best way to unlock the Power of the Force?
* How is the power of the Force best used?

⇕ The Battle between Good and Evil: Jedi Knights and Sith Lords

For several millennia, the Knights of the Jedi Order have been the guardians of peace in the galaxy, with the Force as their ally. The Jedi believe that the path to the Force is through contemplation, passivity, and inner peace. By disciplining themselves to remain calm no matter what is happening around them, Jedi can access and use the power of the Force.

The Jedi regard the power of the Force with respect and responsibility, specifically:

* A Jedi views the Force as a power to merge with in order to further the cause of good.
* Jedi live by a very strict code that is built on tremendous

discipline, selflessness, and self-control. This is why very few Force adepts who begin training with the Jedi ever actually become Jedi Knights, and fewer still become Jedi Masters.

The Sith, on the other hand, have learned to unlock the power of the Force in a completely different way:

* Rather than accessing the power of the Force through inner peace, the Sith have learned to tap the power of the Force by giving in to extreme emotion.
* By letting feelings like anger or fear take over, the Sith can use emotion as a conduit to channel the power of the Force.

There is a common misconception that the Sith access the Force through evil or that being a Sith Lord automatically means that you are evil. This is not true. Any extreme emotion can access the Force. The essence of the Sith's power is in passion, not evil. Passion can take the form of joy or love as well as anger or hatred. Sith are typically evil in that they are often blinded by their passion or corrupted by the power granted to them by the Force.

This plays into the second major difference between the Jedi and the Sith. The Sith believe the Force is a tool to be used, and that the power in and of itself is justification enough for their actions. In short, they believe they can do anything they want simply because they have the power to do so.

Differences aside, the Jedi and Sith share a common lineage, and the origin of the Sith Order starts in the very halls of the Jedi Council.

⚡ A "Sithistory" Lesson

The history of the Sith starts almost twenty-five millennia before the events of *Star Wars: Episode I*. The Jedi Order was formed during this time, as was the Galactic Republic. Up until that point, the Force was practiced in varied and undisciplined ways. On many worlds the Force was misunderstood and often considered to be magic. Force adepts were either revered and worshipped, or shunned and feared.

The founding members of the Jedi Order were Force-sensitive individuals who understood the power and potential of the Force and also foresaw its abuse and danger. They agreed to pool their knowledge of the Force and continue to study it—and most importantly, to educate and regulate those who possessed the power. The Jedi Order was dedicated to making sure the Force was used to ensure peace in the galaxy.

Though the dark side of the Force was not well understood by the Jedi, they were aware of its presence. Early Jedi had all felt the lure of the dark side and those who eventually pooled their knowledge had agreed that the dark path was not one ever to be tread upon by a Jedi, not even for a moment. However, not all of the early Jedi agreed with that philosophy, as we'll find in the next section.

EARLY SITH INCARNATIONS

The term "Sith" dates back to a rough draft of the *Star Wars* script. In early versions the "Knights of Sith" battled the legendary "Jedi Bendu" warriors. In later drafts Lucas renamed the dark order the "Legions of Lettow." Finally, the order was pared down to a single warrior, Darth Vader, Dark Lord of the Sith.

A Long, Long, Long, Long Time Ago...

The debate about the value of the dark side of the Force was waged tirelessly in the Jedi Council. Dissident Jedi argued that the dark power could be controlled and used to further the Jedi cause, but the council didn't agree. This debate eventually gave way to struggle, then war. This war came to be known as the Great Schism. For one hundred years, Jedi fought Jedi in the first conflict between the light and dark sides of the Force.

Because of greater numbers, the Jedi on the light side of

the Force eventually prevailed and decimated the dark Jedi. The battered, surviving dark Jedi were captured and exiled. They were ordered to take the remnants of their broken fleet across the galaxy into uncharted space and never again darken the doorstep of the Republic.

As they traveled backwater space in exile, the outcast Jedi eventually came across the uncharted world of Korriban. There they found a primitive race of Force adepts known as the Sith. The native Sith knew nothing of the Force and used their untapped power for simple magic. The dark Jedi immediately saw an opportunity. They conquered and enslaved the Sith.

For the next several thousand years, the Sith worshipped the dark Jedi as gods. The rebel Jedi came to call themselves Sith Lords. These Sith Lords used the indigenous people as slave labor to build a great new civilization. Inter-breeding blurred species lines, and eventually the term Sith came to mean the off-world native and hybrid peoples of Korriban.

The Golden Age of the Sith

The dark side of the Force was studied and practiced with the same resources and diligence as the light side of the Force. The Sith grew in numbers and power, eventually spreading to the many systems near Korriban, and the Sith Empire was born. The Sith Empire was led by one all-powerful dark master called the Dark Lord of the Sith. There was only one, and this title was usually won by defeating the previous Dark Lord of the Sith. The first Dark Lord of the Sith, it is worth noting, ruled 5,500 years before the events of *Star Wars: Episode I*. He was Marka Ragnos.

Ragnos defeated the previous Dark Lord, Simmus, for the title, and Ragnos's rule ushered in the pinnacle of the Golden Age of the Sith. Ragnos was in power for nearly two hundred years. He brought unparalleled prosperity, growth, and order to the Sith Empire. Ragnos was never successfully challenged and eventually died of old age. At his funeral two Sith Lords named Naga Sadow and Ludo Kressh began to duel over who would next hold the title Dark Lord of the Sith.

Sadow believed that the Sith needed to expand the Empire, while Kressh believed that the Sith needed to avoid attracting the attention of the Jedi. Both Sith Lords knew of the Jedi and the Republic, but knowledge of the Sith's ancient enemy was not something shared by many other Sith. After remaining hidden for thousands of years, the Sith were unsure if the Republic and the Jedi still existed.

Then two hyperspace explorers from the Republic accidentally crashed on Korriban, unwittingly providing the Sith with the proof and information they needed to declare war.

DON'T TELL DARTH VADER, BUT...

"Sith" is the Gaelic word for a fairy or sprite. Gaelic lore describes the Sith as good-natured, tiny, winged creatures that live in the woods.

⚡The Sith Wars

The Great Schism was only the first of many great and bloody wars among the Sith, the Jedi, and the Galactic Republic. In every other instance, the Jedi would defeat the Sith at great cost. The Sith always managed to reemerge, hundreds, sometimes thousands of years later to start their campaign against the Jedi once again. Here is a brief recap of some of the more significant Sith wars.

DARK LORD OF THE CHIMP

The Emperor first appeared in *The Empire Strikes Back*. He was portrayed by compositing the image of an old woman wearing prosthetic makeup with the superimposed image of chimpanzee eyes. In the special edition of *Empire*, this bizarre visage was replaced with the image of actor Ian McDiarmid who played that character in four other *Star Wars* movies.

⇌ The Great Hyperspace War

In the midst of a temporary truce, the two Dark Lord candidates Naga Sadow and Ludo Kressh convened to discuss the fate of Jori and Gav Deragon, the two hapless explorers who stumbled across the Sith Empire.

Kressh saw the intruders as a threat and wanted them executed immediately, while Sadow saw them as a path back to the Republic and eventual control of the Galaxy. Sadow eventually staged a rescue attempt on Korriban and blamed the Republic. Outraged, the Sith voted Sadow onto the throne of the Dark Lord and authorized Sadow's invasion of the Republic.

The Sith invaded and brutalized the Republic. The Jedi fought hard to defend the Republic, but in the end the leader of the Sith invasion force betrayed Sadow and turned the tide of the war. Sensing imminent defeat, Sadow detonated a nearby star, wiping out billions of Jedi, Sith, Republic forces, and innocent people.

Exiled by the Sith, Sadow took the remainder of his broken fleet and created a new Sith Empire on the remote world of Yavin IV. He ruled there until his death one hundred years later. All that remains of Sadow's Empire are the ruins of the Massassi Temples.

⇌ The Feedon Nadd Uprising

Over the next thousand years, the Sith on Korriban all but died out and Sadow's empire was a distant memory. An ambitious Sith Lord named Feedon Nadd was fed up with the decline of the Sith Empire and wanted to reform it and become the new Dark Lord of the Sith. Unable to defeat the current Dark Lord, Nadd left Korriban and settled on the planet of Onderon. Nadd took over the ruling family of Onderon and ruled until his death. For generations afterward, the power of the dark side was the driving force behind the government of Onderon.

When the Beast Wars erupted on the planet, the Jedi Council sent three Jedi, including Ulic Qel-Droma, to restore peace. Though the Jedi successfully stopped this war, their intervention fueled an uprising. The worshippers of Freedon Nadd revolted, but the Jedi quickly ended this bloody battle.

⇌ The Great Sith War

Four thousand years before the events of *Star Wars: Episode I*, the Sith Empire was extinct. It was born again in the hands of a Jedi named Exar Kun. Fascinated by the dark side, Kun secretly studied the Sith during his Jedi training. His research led him to the moon of Onderon. From there he traveled to the former Sith home world of Korriban and later to

Yavin IV. Exar was led down the dark path by the ghost of Freedon Nadd who revealed all of the Sith's dark secrets to the curious Jedi. Kun was eventually seduced by the dark side and swore his allegiance to the Sith.

At the same time, on the recently war-torn world of Onderon, a secret society of aristocrats called the Krath practiced the rituals of the Sith right under the noses of the Jedi.

The Jedi Ulic Qel-Droma discovered the Krath and set out to stop them. In retaliation, the Krath had Ulic's master killed by assassins. Ulic vowed to destroy the Krath from the inside, so he faced them and joined the dark organization. He was not strong enough to resist, however, and Ulic eventually slipped to the dark side. As a Jedi, Ulic was the Krath's greatest instrument.

Eventually the power of two Sith amulets—one possessed by Exar and one by Ulic—led the two Jedi to one another. The two dark Jedi were ready to kill each other when they were visited by a council of ghosts made up of past Dark Lords. The shades instructed them to work together to resurrect the Sith Order and restore the former glory of the Golden Age of the Sith. With Exar Kun as the Dark Lord of the Sith and Ulic Qel-Droma as his apprentice, the two Sith Lords waged the bloodiest war in Galactic history—the Great Sith War.

Exar Kun returned to the Jedi Academy and began luring students to the dark side. Ulic took a small group of ships and used dark-side trickery to begin capturing as many ships as he could. He was to build the next great Sith fleet. In his travels Ulic came across the savage Mandolorian warriors as they attacked a peaceful planet. Ulic intervened on behalf of the planet, but only in order to arrange a one-on-one battle to the death with the Mandolorian leader. Ulic bested Mandolore in hand-to-hand combat and then took control of the Mandolorian hordes, adding them to his ever-growing battle force.

Exar and Ulic allowed the Republic to learn of their invasion and lured the Republic fleet away from Coruscant. The Sith forces then attacked the Capitol world. Ulic was captured by the Jedi but was freed at his trial by Exar. Exar killed his former Jedi master that day and then compelled other Jedi students to do the same.

Later in the war, Ulic unleashed a powerful Sith weapon that caused a chain reaction and destroyed many stars. The fallout from this chain reaction vaporized many systems, destroying the Sith fleet and killing billions and billions of innocent people.

In the final days of the war on the planet Ossus, Ulic faced and killed his own brother. Horrified at what he had become, Ulic fell apart and was defeated by a Jedi who used a forbidden Force-block spell that destroyed Ulic's ability to use the Force forever.

The Republic eventually tracked Exar to Yavin IV and began orbital bombardment of the planet. Surrounded by the mutated descendants of Sadow's followers, Exar used powerful Sith magic to drain these followers' life force and seal his own in the walls of the Massassi Temple. Exar Kun's body was destroyed, but his evil spirit still remained.

⇹ The Jedi Civil War

Shortly after the Great Sith War, the remnants of the Mandalorians sensed weakness in the Republic and decided to strike. A long and bloody war followed between the Jedi and the Mandolorians, but in the end the Jedi prevailed. Two Jedi named Revan and Malak led a Republic fleet to the outer rim and successfully wiped out the Mandalorians. At the end of the war, Revan and Malak were revered as heroes.

Little did the Republic know that their Jedi heroes were destined to become two of the most brutal Sith Lords ever known. While in deep space, the two Jedi landed on Korriban and uncovered many secrets of the Sith. Corrupted by their newfound knowledge and the brutality of the war they had just won, the two Jedi renounced the Order, became Sith, and changed their names to Darth Revan and Darth Malak—master and apprentice.

The two Sith Lords had the loyalty of Jedi and Republic forces that they led to destroy the Mandalorians. They took their new Sith fleet and waged a two-year war on the Republic's outer-rim territories. Eventually, power-hungry Darth Malak tried to destroy both his master Revan and the Jedi. His plan failed, but it did allow the Jedi to capture Revan alive.

The Jedi wiped the Sith master's memory, brainwashing and retraining Revan as a Jedi. In a most uncharacteristic move, the Jedi Council approved the use of Revan as a pawn to find the fugitive Sith Lord Darth Malak. Eventually Revan tracked Malak to Star Forge and defeated him in a lightsaber duel. After that Revan disappeared. The Sith were once again vanquished.

⚡ The New Sith War

Just as they had been born twenty-five thousand years ago, the Sith were born again in the halls of the Jedi Council. This happened one thousand years before the events of *Star Wars: Episode I*.

It began when an unknown, lackluster Jedi began to discover the power of the dark side. He was shunned by the Council and forbidden to continue his study of the dark side. He eventually left the Order and was able to take other Jedi with him, convincing them that the Council was holding them back. After a short while and a few more defections, the new Sith order had grown to fifty sworn Lords.

As time passed, the new Sith Order recruited an army of minions and declared war on the Jedi Order. The war was brutal but short, played out over several small battles. In the end, the Sith were responsible for their own defeat. The infighting, poor organization, and endless power struggles destroyed the Sith Order from the inside.

The final battle of the New Sith War took place on the planet of Ruusan. The Jedi backed the Sith Lords into a cavern and were only minutes away from ending the war. In the final moments of the battle, the Sith Lord Darth Bane was arguing with another Sith Lord about the best way to turn the tide of the battle. He had a plan, but so did the others.

Bane was betrayed in the heat of battle by a Sith Lord named Kaan, and this resulted in the massacre on Ruusan. Kaan detonated a bomb that killed all the Jedi and all the Sith Lords save one, Darth Bane. The Jedi felt they had finally wiped out the Sith. They were careful to guard all of their secrets and closely monitored students who showed dark-side tendencies. After thousands of years of war and mistakes, the Jedi were determined to never again let the Sith rise.

Darth Bane, the sole surviving Sith Lord, slipped away into the shadows. Bane reformed the Sith doctrine and founded what is known as the modern era of the Sith. Now there are only two Sith at a time—a master and an apprentice. They bide their time, work from the shadows, pass the Sith tradition down from master to apprentice, and wait for the day they will have their revenge.

⇟ The Clone Wars

That revenge finally came almost two hundred years later, when a senator from the peaceful planet of Naboo rose to power as Chancellor of the Republic. Senator Palpatine orchestrated an invasion of his home world of Naboo, thus setting a series of deft political moves into motion. No one around him knew that Chancellor Palpatine was actually Darth Sidious, Dark Lord of the Sith. The Republic was now controlled by the Sith.

Sidious's first apprentice, Darth Maul, was killed in the invasion of Naboo by the Jedi Obi-Wan Kenobi. He later took a second apprentice, Darth Tyranus (a well-respected Jedi and political idealist named Count Dooku), whom he used to lead the orchestrated Separatist movement that threatened the Republic with war.

Darth Sidious used the threat of war to justify the creation of a huge clone army. The Clone Wars had begun.

The Jedi and clone army fought side by side, eventually defeating the Separatists. Darth Sidious blamed the Clone Wars on the corruption of the Republic and used the special emergency powers granted to him during the wars to name himself Emperor.

⇟ The Jedi Purge

The Jedi Council was opposed to Emperor Palpatine (Darth Sidious). The Emperor named the Jedi enemies of the

Republic and used one of the Clone Wars' greatest heroes, a Jedi named Anakin Skywalker, to lead the clone army against the Jedi. Anakin killed Darth Tyranus, went over to the dark side, and became Darth Vader, the most legendary of all Sith Lords.

Vader and his clone army began to hunt down and destroy all of the Jedi. Vader eventually faced his former Jedi master, Obi-Wan Kenobi, in a lightsaber battle. Kenobi defeated him and Vader was seriously wounded. He was restored by several operations that made him half man, half machine. Now in his new form, Darth Vader became the dark enforcer of the Emperor. Sidious dissolved the Republic, founded the Empire and, for the first time ever, secured the Sith as the rulers of the galaxy.

The Galactic Civil War

For twenty years the Empire ruled with an iron fist. The Jedi and the Force were long lost, but a tiny rebellion of politicians and warriors formed to oppose the rule of the Emperor. The first great battle of the Galactic Civil War was the Battle of Yavin. Rebel pilot Luke Skywalker made a force-guided shot that destroyed the Empire's mighty battle station, the Death Star. Darth Vader managed to escape the battle with his life and kill his former master Obi-Wan Kenobi.

After the Battle of Yavin, Vader took the Imperial fleet

and scoured the galaxy, searching for the rebel base and more importantly his son, Luke Skywalker. Both Vader and the Emperor felt that Skywalker was a great threat but could be a powerful ally if he could be turned into a Sith Lord. Luke Skywalker had actually been training with Jedi Master Yoda, who hid for years on the backwater planet of Dagobah.

After Vader and his forces crushed the rebels in the Battle of Hoth, he continued to search for his son. They eventually faced each other in a saber duel in Cloud City. After wounding Skywalker, Vader pleaded with son to join him. Vader wanted to destroy the Emperor, end the Civil War, and rule the Galaxy together with his son as his apprentice. Luke refused and escaped.

The Empire built a second Death Star, and the Rebel Alliance set out to destroy it. Skywalker knew his destiny was to face his father. He believed that his role was not to destroy his father but to turn him back to the light side of the Force. Luke surrendered to Darth Vader and was taken before the Emperor.

The final battle of the Galactic Civil War was fought on many fronts. The Emperor lured the rebels to the second Death Star and trapped their fleet between his own fleet and the functional, though incomplete, battle station. On the forest moon of Endor, a band of rebels fought to knock out the shield generator that protected the second Death Star.

In the Emperor's throne room, Jedi Knight Luke Skywalker fought for his destiny, as Darth Sidious tried to turn him to the dark side of the Force.

Luke eventually faced his father again in a lightsaber duel. After a moment of weakness revealed Luke's knowledge of his sister, Leia, to Vader, the Dark Lord promised to turn her to the dark side. Giving in to the dark side, Luke attacked his father and wounded him. Then he regained control and refused to fight. The Emperor, disgusted with Luke, began killing him with Force lightning.

Realizing his life had been spared by his son, Vader turned on his master and killed the Emperor. Vader died minutes later, after reconciling with his son.

The rebels took down the shield generator and destroyed the Death Star. Both the war and the Empire were over.

NAME CHANGE

In early drafts of the *Star Wars* script the Emperor was named Cos Dashit (we're not kidding). Dashit ruled from the Imperial throne world of Alderaan. In later drafts, he was simply referred to as the Emperor.

⚡ The Sith Will Never Die

After the Battle of Endor, the Sith continued to cause problems for the fledgling New Jedi Order established by Jedi Master Luke Skywalker. Using the deepest and darkest of the Sith arts, the ghost of Emperor Palpatine (Darth Sidious) eventually found its way to a new clone body. In an attempt to stop the new Emperor, Luke traveled to face him on the dark world of Byss. This resurrected Sith Lord managed to turn Skywalker to the dark side, but he was later saved by his Jedi sister, Princess Leia Solo.

Later, the ghost of ancient Sith Lord Exar Kun was awakened from its sleep deep within the Temples of Yavin IV when Luke set out to establish a new Jedi Academy in the old ruins. Kun's spirit corrupted two of Luke's students and trapped the Jedi Master in a spirit realm before Luke and his students were able to destroy the specter. They did this by using the light side of the Force to fight Kun in both the corporeal and spirit realms.

It's clear that as long as there are Jedi in the galaxy, there will also be the potential for the Sith to rise again.

CONTROLLING CHAOS: HOW THE DEATH STAR WORKS

There is no question that the last few years of world history have seen a fair share of chaos and disorder. With so much global unrest, governments worldwide are struggling to devise new methods to maintain order. The Galactic Empire's solution to order is the Death Star.

Simply put, the Death Star is the most ambitious spacestation project in Galactic history. It takes more than a million military personnel to run this station, which is the size of a small moon.

But the Death Star is no moon. Despite its size, the Death Star can still travel at superluminal velocities, just like spacecraft with only a fraction of its mass. But the real power and purpose of the Death Star lies in its superlaser. The Death Star's superlaser is capable of firing a beam of directed energy powerful enough to completely destroy a planet of any size! Needless to say, nothing quells chaos and disorder faster than the threat of total planetary annihilation.

The Death Star represents the absolute pinnacle of military engineering and technology. In this section,

HowStuffWorks will look at the Death Star inside and out, examine the fascinating history behind this powerful military and political tool, discover other incarnations of the Death Star, and learn about what really happens when you blow up a planet.

⚡ Star Wars 101: Death Star Basics

The Death Star's main purpose is to function as a mobile platform for its main weapon—the superlaser. The Death Star's structure is basically an enormous housing for the superlaser and the reactor that powers it.

It takes a lot of technicians to operate and maintain the superlaser. And although it's the most powerful weapon in the galaxy, the superlaser is completely defenseless if attacked directly. So the superlaser also needs military support to defend it. To address these issues, the designers of the Death Star equipped this enormous housing to serve two purposes: it is both a mobile weapons platform and a fully operational battle station.

Of course, in order for the Death Star to be a real threat, it has to be mobile. To accomplish this, the Death Star features a complex network of real-space ion engines and hyperdrive field generators that allow it to travel like any other interstellar spacecraft.

Basically the Death Star is made of four major parts:

- The battle station.
- The superlaser.
- The propulsion system.
- The hypermatter reactor that powers it all.

Let's look at the science behind all of these crazy components.

⚡The Tip of the Iceberg: The Death Star Surface

In the original design, the Death Star measures 120 kilometers (roughly 75 miles) in diameter. A huge equatorial trench splits the surface of the station into two hemispheres. This trench is used to house many of the Death Star's main systems:

- Landing bays.
- Drive thrusters.
- Sensor arrays.
- Tractor beam systems.

In addition to the main trench there are two supplementary trenches halfway between the equator and each pole that are used mostly for maintenance and secondary reactor ventilation.

Thousands of weapons emplacements are scattered all across the Death Star's surface, including:

* 10,000 turbolaser batteries.
* 2,500 laser cannons.
* 2,500 ion cannons.
* 768 tractor-beam projectors.

Most of the surface of the Death Star is covered with buildings of varying size and purpose, so that it closely resembles a sprawling metropolis. Now let's look inside.

Inside the Death Star

Most of the space inside the Death Star is devoted to systems required to maintain the superlaser, propulsion system, and hypermatter reactor. Of course the largest space is the main reactor chamber at the core of the Death Star. The rest of the interior is made up of a honeycomb of decks for personnel and equipment. This space is designed with two separate layouts, each with a different source and orientation of artificial gravity.

The "layer" closest to the surface is laid out in a series of concentric decks with artificial gravity generators pointing toward the Death Star's core. Below this are thousands of levels of sprawling stacked decks dotted with vast, deep shafts that all link to the reactor's main chamber. This section of the Death Star makes up the bulk of the interior and has gravity pointing toward the station's southern pole.

The two interior sections of the Death Star are divided into twenty-four zones, twelve per hemisphere. Each zone is organized into six sectors:

- General.
- Command.
- Military.
- Security.
- Service.
- Technical.

Each of the sectors is run by an officer who answers to a zone captain who controls his or her zone from a zone bridge. The entire Death Star command structure answers to one Death Star Commander. The Commander operates from the overbridge. The overbridge is the nerve center of the Death Star and is located just above the top edge of the superlaser dish. Governor Tarkin was acting commander of the first Death Star. The station's military forces fell under the command of General Tagge (ground forces) and Admiral Motti (naval forces).

"WHAT A WONDERFUL SMELL YOU'VE DISCOVERED."

When a small group of Rebels infiltrated the first Death Star to rescue Princess Leia, they avoided capture by hiding in a conveniently located trash compactor on the detention level. Over the years many questions have been raised about that trash compactor. Why was the trash compactor there? What was that creature in there and how did it get in there in the first place? All of these questions and more are addressed in Joshua Tyree's insightful essay, "On the Implausibility of the Death Star's Trash Compactor" on McSweeneys.net. We recommend checking it out.

⚡ The Superlaser

If you've ever burned a leaf with a magnifying glass, you understand the basic principle behind the superlaser. When a magnifying glass is held at the correct angle between the sun and a leaf, the sun's rays are focused through the lens. These rays intersect under the lens, and at the point of intersection, a beam of heat is created that burns the leaf. The sun is the source of power and the lens is the focus.

The superlaser has a massive lens built around a huge

synthetic focusing crystal. The lens is known as "the Eye" and is surrounded by eight tributary lasers. There are also four backup lasers in case any of the main eight tributaries fail. All of the tributary lasers can be angled for targeting. This allows the Death Star to aim the superlaser within a certain field of fire without having to turn the entire station. The main cannon and eight tributary lasers fire beams that converge at the outer perimeter of the superlaser dish in an amplification nexus. A main beam then blasts from the nexus to the intended target.

The Death Star's superlaser derives power directly from the hypermatter reactor. The lasers convert and focus the full power of the reactor to create the beams. So going back to our magnifying glass example, the superlaser is like a series of large magnifying glasses focusing the entire power of the reactor (which is like a small sun) into one huge beam to destroy a planet, rather than a few rays of light to burn a leaf.

⚡Firing the Superlaser

There have only been two instances where the Death Star fired its superlaser at full power while targeting a planetary body. The Death Star's first test firing destroyed the remote and uninhabited planet Despayre. The first Death Star was built in orbit around Despayre, making the planet an ideal choice to test the superlaser's power as well as destroy the evidence of the Death Star's construction. The second instance was the highly publicized destruction of Alderaan.

The superlaser's power needs to be recharged between blasts, limiting it to only one planet-destroying beam per day. However, the output of the superlaser can be scaled to fire at smaller targets, such as capital ships. The superlaser can produce a scaled beam at a recharge rate of one per minute.

It takes 168 Imperial gunners to operate the superlaser. Fourteen gunners man each tributary laser while the remaining crew operates other systems. The superlaser can only be fired under direct orders from the station commander or the Emperor himself.

NICE TRY

There was an ill-fated attempt to fit a smaller axial-superlaser variant on Eclipse- and Sovereign-class Star Destroyers. These superlasers were not capable of destroying a planet but could sear the surface and boil away atmosphere as well as large bodies of water. The project was never a success.

⚡Power and Propulsion

The greatest challenge in designing the Death Star was not creating a cannon big enough to fire a beam that could destroy a planet, nor was it creating a battle station the size of a small moon. The greatest challenge was always powering a cannon big enough to fire a beam that could destroy a planet and moving a battle station the size of a small moon. The answer to both of these problems was solved with the invention of the hypermatter reactor.

The hypermatter reactor is the heart of the Death Star. The Death Star's hypermatter core is based largely on the early Sienar Systems hypermatter implosion core that was the power source of the Confederacy of Independent Systems' Great Weapon (the early inspiration for the Death Star— more on this later). Little is actually known about the details of the highly classified reactor design, but we do know that it is a massive fusion reactor fed by stellar fuel bottles that line the periphery of the main reactor chamber.

The Death Star's real-space propulsion system is made up of a network of ion engines that use converters to transform reactor power into thrust. The engine thrusters are primarily lined along the equator of the station.

Hyperspace travel is made possible with linked banks of hyperdrive field generators. Each bank contains 123 hyperdrive field generators. They are all tied together into one navigational matrix that is controlled from the overbridge.

"THAT'S NO MOON."

On July 3, 2004, soon after orbital insertion around Saturn, the unmanned spacecraft Cassini sent back a disturbing image back to Earth.

While many believed it to be proof of the existence of a new Death Star in our own solar system, it turns out that's no space station. It was a picture of Mimas, one of the thirty-one moons of Saturn. It's 247 miles across (398 kilometers) and sports a huge crater named Herschel that is 80 miles wide (130 kilometers). The picture was taken by the Cassini spacecraft from about a million miles (1.7 million kilometers) away!

⚡Life on the Death Star

It takes more than a million people to operate the Death Star, and there is room for over a billion people on board. At least 1,161,293 Imperials are stationed on the Death Star at any given time. The standard complement of personnel includes:

- 265,675 station crew.
- 52,276 gunners.
- 607,360 troops.
- 25,984 stormtroopers.

* 42,782 ship support staff.
* 167,216 pilots and support crew.

The station also carries:

* 7,200 starfighters.
* 4 strike cruisers.
* 3,600 assault shuttles.
* 1,400 AT-ATs.
* 1,400 AT-STs.
* 1,860 drop ships.
* A variable number of support, recon, and assault droids.

Tours on the Death Star last at least 180 days, and usually much longer. Personnel are often in deep space without leave for months at a time, and since the location of the Death Star is always classified, contact with family or friends is strictly prohibited. This can make life on the Death Star very difficult. To ease the burden of this duty, the station is outfitted with many civilian amenities. The general sector of each zone in the Death Star has a park, shopping centers, and recreation areas that include restaurants, a cinema, and fitness centers.

REVISIONIST HISTORY?

The first Death Star was destroyed in the Battle of Yavin and everyone aboard was lost. Though the Rebel Alliance official reports put the death toll at around a million Imperials, that estimate is based on intelligence about the minimum crew requirements to operate the Death Star. The Imperials tell a different story.

Imperial analysts claim that between 800 million and a billion people were lost in the "Yavin Massacre." This number is based on the flood of missing personnel reports filed in a period of eighteen months after the Battle the Yavin. Imperial loyalists claim that the Alliance intentionally downplayed the loss of life to distract from the fact that the destruction of the Death Star was as catastrophic a loss of life as the destruction of Alderaan. The crew manifest of the Death Star was classified and destroyed with the station, so the truth may never be known.

A Doctrine of Fear: The History of the Death Star

The Death Star battle station was the brainchild of Imperial bigwig Grand Moff Tarkin. It was the centerpiece of Tarkin's

Doctrine of Fear that created sweeping reform in the structure of the Empire. It was also the final step in solidifying the Emperor's total authority.

The largest change to the Empire made by the Doctrine of Fear was the dissolution of the Imperial Senate. Tarkin's policy put power directly in the hands of the regional governors. The governors, who ruled over several planetary systems, now answered directly to the Emperor under the Doctrine of Fear. This new organization greatly reduced the bureaucracy in the Empire and greatly increased the power of regional governors like Tarkin.

Opponents of the Doctrine of Fear (and there were many in the Empire) claimed that this new policy would rip the Empire apart and that planets would revolt without direct representation in the Senate. Tarkin's answer to any potential dissonance was the Death Star.

Tarkin intended to make an example of a rebellious system as soon as the Death Star was capable. He believed that after the total annihilation of a planet, fear would spread throughout the galaxy, making order easy to maintain. The development and construction of the Death Star had begun long before the first debate about the Doctrine of Fear ever took place. In fact, the Doctrine of Fear was ratified by the Emperor the very same day the second successful firing of the Death Star destroyed the planet of Alderaan.

Visionary scientists and engineers like Qwi Xux, Tol Sivron, and Bevel Lemelisk were conscripted by the Empire to develop the space station. They all worked in a top-secret facility, code-named Maw, hidden deep in one of the most inhospitable regions of the galaxy. A prototype Death Star was built at Maw but was little more than a spherical frame with an untested superlaser. It was much smaller than the Death Star would eventually be and had no propulsion system.

The construction phase of the Death Star project took place in orbit around the planet Despayre and was primarily handled by defense contracting company Sienar Systems. Sienar had worked on a prototype of a similar space station many years earlier for the CIS and was contracted by Tarkin to do work on the Death Star project. (It should also be noted that Sienar Systems CEO Raith Sienar was "coincidentally" an old friend of Grand Moff Tarkin). For years, the Empire used prison labor to mine Despayre for materials. Prison laborers also handled the more menial and dangerous aspects of the station's construction.

THE GREAT WEAPON

The inspiration for the Death Star came from a Separatist super-weapon called the Great Weapon. The Great Weapon was a moon-sized space station with a large laser cannon developed by the Trade Federation, Geonosians, and Techno-Union for use in their war against the Old Republic. It was never put to use and was captured by the newly formed Empire after the Clone Wars. The Great Weapon was never completed.

The original Death Star was destroyed in the Battle of Yavin by rebel pilot Luke Skywalker. Skywalker fired a proton torpedo into a reactor vent shaft. The ensuing chain reaction caused a critical overload of the reactor, destroying the entire station and killing everyone aboard.

⚡Reincarnated: Death Star II

After the destruction of the first Death Star, the Empire immediately began building a second Death Star. Death Star II was placed under the command of Moff Jerjerrod and was a larger (160 km/99.5 miles in diameter), more powerful version of its predecessor.

Many of the design flaws of the first Death Star were corrected in Death Star II, including its Achilles heel—the thermal exhaust ports. Instead of relying on several large vent ports that led directly to the station's main reactor, Death Star

II was designed with a network of variably angled millimeter-wide heat dispersion ducts. This design made it impossible for a projectile to reach the reactor via the vent system.

Another costly oversight was corrected in the layout of the new station's defenses. The original Death Star was only designed to repel an attack from large capital ships. Because of the "loose" net cast to catch attackers, rebel fighters were able to easily penetrate the Death Star's defense grid. The design of Death Star II almost doubled its surface defenses and triangulated them so that both capitol ships and fighters could easily be gunned down.

Death Star II was designed with:

- 30,000 turbolaser batteries.
- 7,500 laser cannons.
- 5,000 ion cannons.

In addition, Death Star II featured a larger and more powerful superlaser that was capable of firing more frequently and more accurately.

Death Star II was under construction in orbit around the forest moon of Endor. While under construction, the new Death Star was protected by a large energy shield projected from the nearby moon. Despite this, the Rebel Alliance destroyed the unfinished Death Star II in the Battle of Endor.

⚡So What Happens If You Blow Up a Planet?

There has been a lot of controversy surrounding the Death Star. Besides the obvious issues associated with destroying an entire planet, there are also concerns about the effects that the destruction of a planet would have on other planets in the same system.

So what does happen to other planets in the same system as a planet that is completely destroyed by the Death Star?

To answer this question HowStuffWorks went to F. Todd Baker, physics professor at the University of Georgia. Here was his answer:

Regarding the moons of a planet, the subsequent motion would be determined by what happens to the debris from the planet. Of course the "destruction" of a planet does not mean that its mass disappears, it is just redistributed. A couple of scenarios:

1. Suppose Earth became a cloud of debris with ten times the radius of the present Earth. Then this cloud would continue orbiting the sun as it does now (the length of a year would be the same) and the moon would continue orbiting this cloud as it does the earth now (the length of a month would be the same). This supposes that the cloud of debris were roughly spherically symmetric.

2. Suppose that the annihilation as so catastrophic that the debris completely dispersed in a spherically symmetric way. As soon as some of the debris passed the moon's orbit, the gravitational force on the moon would begin decreasing and the moon would change its orbit in a continuous way until, finally, it would be orbiting the sun in an orbit more or less the same as Earth's current orbit. All the debris would also end up orbiting the sun in many different kinds of orbits, much as asteroids and comets do today. Some would end up in orbits that resulted in their hitting the sun.

In the next section, we will look at some of the other great technologies we've seen in the cosmos from the great pioneers of fictional space in *Star Trek*.

MAY THE FORCE BE WITH YOU: A STRIKING *STAR WARS* QUIZ

We've thrown a lot at you in our intergalactic exploration of the *Star Wars* super-universe. Just to make sure you're on your guard, let's test your knowledge of all things *Star Wars* with this quiz. (Note: Using the Force is considered cheating.)

Check your answers in the back of the book on page 278.

1. **True or False: The Sith always act in groups to increase the force of their attack.**

    a. True

    b. False

2. **What is the name of the remote and uninhabited planet that was destroyed by the first test of the Death Star's laser cannon?**

    a. Despayre

    b. Alderaan

    c. Tatooine

3. In the *Star Wars* series, what serves as the energy supply for a lightsaber?

   a. Crystal cell

   b. Phoenix feather

   c. Diatium core

4. Who did Anakin Skywalker have to defeat to become the most legendary of all Sith Lords?

   a. Darth Tyranus

   b. Darth Sidious

   c. Darth Vader

5. True or False: According to *Star Wars*, not just anyone can tap into the Force and use it at their will.

   a. True

   b. False

6. The Rebel Alliance destroyed the Death Star for the first time during what battle?

   a. Battle of Yavin

   b. Battle of Utapau

   c. Battle of Geonosis

7. Which of the following is not one of the six sectors within each of the zones on the Death Star?

   a. Service

   b. Operations

   c. Technical

8. True or False: The superlaser's power needs to be recharged between blasts.

   a. True

   b. False

9. The Great Schism produced what distinct group?

   a. The Ewoks

   b. The Sith Lords

   c. The Rebel Alliance

10. How did the Rebels finally defeat the Sith in the Galactic Civil War?

    a. The Jedi overwhelmed the Sith with their vast numbers and better lightsaber skills.

    b. The Rebels took down the shield generator and destroyed the Death Star.

    c. The Rebels discovered the home planet of the Sith and destroyed it.

ALL ABOARD THE STARSHIP *ENTERPRISE*: HOW *STAR TREK*'S WARP SPEED WORKS

Onboard the Starship *Enterprise*, you're hanging out with the crew members, enjoying a game of poker. You're traveling at impulse speed during a leisurely deep-space exploration, and everyone has some downtime. But wait—all of a sudden, the ship receives an urgent message from a Federation admiral, informing the crew of an outbreak of war in the Neutral Zone.

The *Enterprise* is ordered to report to the situation as soon as possible. The area in question is about twenty light-years (117 trillion kilometers) away, but this is no problem in the *Star Trek* universe. The captain adjusts the ship's warp drive appropriately, and you settle in for warp speed. Traveling faster than the speed of light, you should arrive to your destination in just a few minutes.

Everything from H.G. Wells's *The Time Machine* to *Star Trek* to Joss Whedon's *Firefly* series has touched on the possibilities of time travel, teleportation, and, of course, warp speed. But how does something like warp speed fit into reality and our universe? Is warp speed just a wacky science-fiction device, or is it theoretically possible? How does it work in the *Star Trek* universe? For everything on warp speed from here to infinity and beyond, keep reading!

⇄ An Object in Motion: Newton's Third Law of Motion

When the writers of *Star Trek* sat down to plan the series, they found themselves confronted with a few problems. They were essentially creating a space opera, a subgenre of science fiction that takes place in space and covers the span of several galaxies and millions of light-years. The *Star Wars* films are another example of the space opera subgenre. As the "opera" part of the name suggests, a show like *Star Trek* isn't meant to be slow or ordinary. When people think of the series, they probably think of melodramatic plots involving aliens, space travel, and action-packed laser fights.

So the creator of the series, Gene Roddenberry, and the other writers had to find a way to move the show's characters around the universe in a timely, dramatic fashion. At the same time, they wanted to do their best to stick to the laws of physics. The biggest problem was that even if a starship could travel at the speed of light, the time to go from one galaxy to another could still take hundreds, maybe thousands of years. A journey from Earth to the center of our galaxy, for example, would take about twenty-five thousand years if you were to travel just under the speed of light. This, of course, wouldn't make very exciting television.

The invention of warp speed solved the opera part of the problem, since it allowed the *Enterprise* to go much faster than the speed of light. But what was the explanation? How could they explain an object traveling faster than the speed of light, something Einstein proved impossible in his special theory of relativity?

The first obstacle the writers had to confront is much simpler than you'd think. One of the most important things you need to know before understanding warp speed is actually one of the oldest tricks in the physics book, Newton's third law of motion. You've probably heard it before—this law states that for every action, there is an equal and opposite reaction. It simply means that for every interaction between two objects, a pair of forces is working on both of them. For instance, if you roll one billiard ball straight into another one that's at rest, they will each exert an equal force on one other. The moving ball will hit the ball at rest and push it away, but it will also be pushed back by the latter.

NEWTON'S LAW IN REAL LIFE

You feel this law come into play every time you accelerate in a car or fly in an airplane. As the vehicle speeds up and moves forward, you feel pressure on your seat. The seat is pushing on you, but you're also exerting a force against the seat.

So what does this have to do with *Star Trek* and the *Enterprise*? Even if it were possible to accelerate to something like half the speed of light, such intense acceleration would kill a person by smashing him against his seat. Even though he'd be pushing back with an equal and opposite force, his mass compared to the starship is just too small. The same kind of thing happens when a mosquito hits your windshield and splatters. So how can the *Enterprise* possibly go faster than the speed of light without killing the members on board?

In the next section, we'll see how the *Star Trek* creators began to get around the problem of sending matter through space at superluminal speeds.

Einstein, Relativity, and the Space-Time Continuum

To sidestep the issue of Newton's third law of motion and the impossibility of matter traveling faster than the speed of light, we can look to Einstein and the relationship between space and time. Taken together, space (consisting of three dimensions: up-down, left-right, and forward-backward) and time are part of what's called the space-time continuum.

It's important to understand Einstein's work on the space-time continuum and how it relates to the *Enterprise* traveling through space. In his special theory of relativity, Einstein states two postulates:

1. The speed of light (about 300 million meters per second) is the same for all observers, whether or not they're moving.

2. Anyone moving at a constant speed should observe the same physical laws.

Putting these two ideas together, Einstein realized that space and time are relative—an object in motion actually experiences time at a slower rate than one at rest. Although this may seem absurd to us, we travel incredibly slowly when compared to the speed of light, so we don't notice the hands on our watches ticking slower when we're running or traveling on an airplane.

TIME LAG

Scientists have actually proved the phenomenon Einstein describes by sending atomic clocks up with

high-speed rocket ships. They returned to Earth slightly
behind the clocks on the ground.

What does this mean for the Captain Kirk and his team?
As an object gets closer and closer to the speed of light, that
object actually experiences time at a significantly slower rate.
If the *Enterprise* were traveling safely at close to the speed
of light to the center of our galaxy from Earth, it would
take twenty-five thousand years of Earth time. For the crew,
however, the trip would probably only take ten years.

Although that time frame might be possible for the individuals on board, we're presented with yet another problem—a
Federation attempting to run an intergalactic civilization would
run into some problems if it took fifty thousand years for a
starship to hit the center of our galaxy and come back.

So the *Enterprise* has to avoid the speed of light to keep
the passengers onboard in sync with Federation time. At
the same time, the *Enterprise* must reach speeds faster than
that of light to move around the universe in an efficient
manner. Unfortunately, as Einstein states in his special
theory of relativity, nothing is faster than the speed of light.
Space travel therefore would be impossible if we're looking
at the special relativity.

That's why we need to look at Einstein's later theory, the general theory of relativity, which describes how gravity affects the shape of space and flow of time. Imagine a stretched-out sheet. If you place a bowling ball in the middle of the sheet, the sheet will warp as the weight of the ball pushes down on it. If you place a baseball on the same sheet, it will roll toward the bowling ball. This is a simple design, and space doesn't act like a two-dimensional bed sheet, but it can be applied to something like our solar system. More massive objects like our sun can warp space and affect the orbits of the surrounding planets. The planets don't fall into the sun, of course, because of the high speeds at which they travel.

In the next section, we'll see how this comes into play on the *Enterprise*.

⚡Pedal to the Metal: The Warp Drive

The ability to manipulate space is the most important concept in regard to warp speed. If the *Enterprise* could warp the space-time continuum by expanding the area behind it and contracting the area in front, the crew could avoid going the speed of light. As long as it creates its own gravitational field, the starship could travel locally at very slow velocities, therefore avoiding the pitfalls of Newton's third law of motion and keeping clocks in sync with its launch site and destination. The ship isn't really traveling at a "speed," per se—it's more like it's pulling its destination toward it while pushing its starting point back.

SPECULATING ON THE FUTURE

Because the ideas behind Einstein's general theory of relativity are complex and still open to interpretation, they leave many opportunities available for science fiction writers. We may not know how to bend space and time with our current technology, but a fictional civilization set in the future may be completely capable of inventing such a device with the right imagination.

In the *Star Trek* universe, warp speed is accomplished through the use of a warp drive. The warp drive is powered by matter-antimatter reactions, which are regulated by a substance called dilithium. This reaction creates highly energetic plasma known as electro-plasma, a type of matter with its own magnetic field that reacts with the starship's warp coils. The warp coils are typically enclosed in what the *Star Trek* writers call a "warp nacelle." The whole package creates a "warp field" or "bubble" around the *Enterprise*, allowing the ship and its crew to remain safe while space manipulates around them.

Sometime between the first television series (*Star Trek: The Original Series*) and the second (*Star Trek: The Next Generation*), the writers decided to assign a limit to warp speed—using a scale of Warp-1 to Warp-10. The *Enterprise* wasn't allowed to travel just anywhere at any time, seeing as that would make plotting too easy. In the show, Warp-10 became an impossible maximum speed, an infinity in which the starship would be at all points in the universe at the same time. Warp-9.6, according to the *Next Generation* technical manual, is the highest attainable speed allowed—it's set at 1,909 times the speed of light. Although there are some inconsistencies, the following list the different speeds in the *Star Trek* universe:

WARP FACTOR	NUMBER OF TIMES THE SPEED OF LIGHT
1	1
2	10
3	39
4	102
5	215
6	392
7	656
8	1,024
9	1,516
9.6	1,909
10	Infinity

In the next section, we'll take a look at some of the problems the concept of warp speed encounters.

⚡Warping Woes: Problems with Warp Speed

So Einstein helped the *Star Trek* writers manipulate space in a science fictional universe, but is it actually possible to build a spaceship that could propel people across vast galaxies in a relatively short period of time?

Physicist Miguel Alcubierre has suggested the use of so-called "exotic matter," a theoretical type of matter with negative energy. If it could be found or created, the exotic matter would do the job of repelling space and time and creating the gravitational field.

Unfortunately, that's as far as it goes for possible fuel sources—there are more problems than solutions when it comes to the concept of powering warp speed. Even if the *Enterprise* were to travel at sub-light speeds, known as impulse drive to *Star Trek* fans, the amount of fuel and energy needed to travel quickly through space would be too much for a single starship.

The impulse drive of the *Enterprise* is powered by nuclear fusion, the same kind of reaction that lights up the sun and creates huge explosions from certain nuclear bombs. According to Lawrence Krauss, PhD, a theoretical physicist and author of *The Physics of Star Trek*, if Captain Kirk wanted to travel at half the speed of light (150,000 kilometers per second), the starship would need to burn eighty-one times its mass in hydrogen, the fuel used for nuclear fusion.

The technical manual for *Star Trek: The Next Generation* lists the *Enterprise* as weighing more than 4 million metric tons, so the ship would need more than 300 million metric tons of hydrogen just to move forward. Of course, to slow down to a stop, the starship would need yet another 300 million metric tons of fuel, and a potential trip across galaxies would require 6,642 times the mass of the *Enterprise*.

Some people have proposed a system in which a device gathers hydrogen as the starship travels, eliminating the necessity to store huge amounts of fuel, but Krauss suggests this device would have to be about twenty-five miles wide to capture anything worth using. Even though hydrogen is the most abundant element in the galaxy, there's only about one atom of hydrogen for every cubic square inch.

FUEL FOR WARP DRIVE

Making the warp drive work would be another thing. The warp drive in *Star Trek* gets its power by reacting matter with antimatter—the result is complete annihilation and the release of pure energy. Since antimatter isn't very common throughout our universe, the Federation would have to produce it, something we can do today at the Fermi National Accelerator Laboratory (Fermilab) in Illinois.

Again, the problem turns out to be an issue of the amount of fuel necessary to power a warp drive. Physicist Lawrence Krauss notes that Fermilab is capable of producing 50 billion antiprotons in one hour—enough to produce 1/1,000 of a watt. You would need 100,000 Fermilabs to power a single light bulb. Producing enough antiprotons to bend the space-time continuum looks nearly impossible using our current technology.

Although there's little chance during this century of humans developing a spaceship that could bend space and travel to distant galaxies faster than the speed of light, this hasn't stopped scientists and fans of the series from thinking about the potential.

THE SCIENCE OF *STAR TREK*

As far as science fiction goes, *Star Trek* is well-regarded by its fans for sticking to relatively plausible physics. Although no one's come up with a starship that would travel at warp speed, no one's disproved the possibility of such a feat. *Star Trek* has also looked at other big concepts throughout the series, including the notion of time travel through black holes or wormholes. The writers also get several nitpicky details correct, such as the fact that there is no sound in space. While George Lucas includes laser blasts and explosions throughout his *Star Wars* series to keep things dramatic, *Star Trek* kept a bit closer to reality by not including sound effects in space.

BEAM ME UP, SCOTTY: HOW TELEPORTATION WILL WORK

Ever since the wheel was invented more than five thousand years ago, people have continued to invent new ways to travel faster and faster from one point to another. The chariot, bicycle, automobile, airplane, and rocket have all been invented to decrease the amount of time we spend getting to our desired destinations. Yet each of these forms of transportation shares the same flaw: they require us to cross a physical distance, which can take anywhere from minutes to many hours, depending on the starting and ending points.

But what if there were a way to get you from your home to the supermarket without having to use your car, or from your backyard to the International Space Station without having to board a spacecraft? Scientists are working right now on such a method of travel, combining properties of telecommunications and transportation to achieve a system called teleportation. In this section, we will learn about experiments that have actually achieved teleportation with photons, and how we might be able to use teleportation to travel anywhere at any time. Cool, huh?

TELEPORTING LIGHT SHOWS

Looking forward to instantaneous travel? The *Star Trek* teleporter is one step closer to reality. Researchers at the Australian National University successfully teleported a laser beam. Could humans be next?

First, what is teleportation? Teleportation involves dematerializing an object at one point and then sending the details of that object's precise atomic configuration to another location, where it will be instantaneously reconstructed in its correct form. What this means is that time and space could be eliminated from travel. We could be transported to any location instantly, without actually crossing a physical distance.

Many of us were introduced to the idea of teleportation and other futuristic technologies *Star Trek: The Original Series*. Viewers watched in amazement as Captain Kirk, Spock, Dr. McCoy, and others beamed down to the planets they encountered on their journeys through the universe.

In 1993, the idea of teleportation moved out of the realm of science fiction and into the world of theoretical possibility. It was then that physicist Charles Bennett and a team of researchers at IBM confirmed that quantum teleportation was possible, but only if the original object being teleported was destroyed.

This revelation, first announced by Bennett at an annual meeting of the American Physical Society in March 1993, was followed by a report on his findings in the March 29, 1993 issue of *Physical Review Letters*. Since that time, experiments using photons have proven that quantum teleportation is in fact possible!

⚡Teleporting Today: Recent Teleportation Experiments

In 1998, physicists at the California Institute of Technology (Caltech), along with two European groups, turned the IBM ideas into reality by successfully teleporting a photon, a particle of energy that carries light. The Caltech group was able to read the atomic structure of a photon, send this information across 3.28 feet (about 1 meter) of coaxial cable, and create a replica of the photon. As predicted, the original photon no longer existed once the replica was made.

In performing the experiment, the Caltech group was able to get around the Heisenberg uncertainty principle, the main barrier for teleportation of objects larger than a photon. This principle states that you cannot simultaneously know the location and the speed of a particle. But if you can't know the position of a particle, then how can you teleport it? To teleport a photon without violating the Heisenberg principle, the Caltech physicists used a phenomenon known as entanglement. In entanglement, at least three photons are needed to achieve quantum teleportation:

* Photon A: The photon to be teleported.
* Photon B: The transporting photon.
* Photon C: The photon that is entangled with photon B.

If researchers tried to look too closely at photon A without entanglement, they would bump it and thereby change it. By entangling photons B and C, researchers can extract some information about photon A. The remaining information would be passed on to B by way of entanglement, and then on to photon C. When researchers apply the information from photon A to photon C, they can create an exact replica of photon A. However, photon A no longer exists as it did before the information was sent to photon C.

In other words, when Captain Kirk beams down to an alien planet, an analysis of his atomic structure is passed through the transporter room to his desired location, where a replica of Kirk is created and the original is destroyed.

The most recent successful teleportation experiment took place on October 4, 2006, at the Niels Bohr Institute in Copenhagen, Denmark. Quantum physicist Eugene Polzik,

PhD, and his team teleported information stored in a laser beam into a cloud of atoms. According to Polzik, "It is one step further because for the first time it involves teleportation between light and matter, two different objects. One is the carrier of information and the other one is the storage medium." The information was teleported about 1.6 feet (half a meter).

Quantum teleportation holds promise for quantum computing. These experiments are important in developing networks that can distribute quantum information. Professor Samuel Braunstein of the University of Wales, Bangor, called such a network a "quantum Internet." This technology may be used one day to build a quantum computer that has data transmission rates many times faster than today's most powerful computers.

Time Travel for Us: Is Human Teleportation in the Cards?

We are years away from the development of a teleportation machine like the transporter room on *Star Trek*'s *Enterprise* spaceship. The laws of physics may even make it impossible to create a transporter that enables a person to be sent instantaneously to another location, which would require travel at the speed of light.

For a person to be transported, a machine would have to be built that can pinpoint and analyze all of the 10^{28} atoms that make up the human body. That's more than a trillion atoms. This machine would then have to send this information to another location where the person's body would be reconstructed with exact precision. Molecules couldn't be even a millimeter out of place, lest the person arrive with some severe neurological or physiological defect.

In the *Star Trek* episodes, and the spin-off series that followed it, teleportation was performed by a machine called a transporter. This was basically a platform that the characters stood on, while Scotty adjusted switches on the transporter-room control boards. The transporter machine then locked on to each atom of each person on the platform and used a transporter carrier wave to transmit those molecules to wherever the crew wanted to go. Viewers watching at home witnessed Captain Kirk and his crew dissolving into a shiny glitter before disappearing, then rematerializing instantly on some distant planet.

If such a machine were possible, it's unlikely that the person being transported would actually be "transported." It would work more like a fax machine—a duplicate of the person would be made at the receiving end, but with much greater precision than a fax machine. But what would happen to the original? One theory suggests that teleportation would combine genetic cloning with digitization.

In this bio-digital cloning, teletravelers would have to

die, in a sense. Their original mind and body would no longer exist. Instead, their atomic structure would be re-created in another location, and digitization would re-create the travelers' memories, emotions, hopes, and dreams. So the travelers would still exist, but they would do so in a new body of the same atomic structure as the original body and programmed with the same information.

But like all technologies, scientists are sure to continue to improve upon the ideas of teleportation, to the point that we may one day be able to avoid such harsh methods. One day, one of your descendants could finish up a work day at a space office above some faraway planet in a galaxy many light-years from Earth, tell his or her wristwatch that it's time to beam home for dinner on planet X below, and sit down at the dinner table as soon as the words leave his mouth.

TRIED AND TRUE TREKKIE TECH: TOP TEN *STAR TREK* TECHNOLOGIES THAT ACTUALLY CAME TRUE

T he transporter we just discussed essentially dematerialized a human body at one point only to rematerialize it in the transporter bay on the ship. Somehow, it broke down atoms and molecules within the body—scattered them through the vacuum of space, from point A to point B, without losing a single one, then voilà, that person re-emerged out of thin air. Sounds pretty cool, albeit impossible, right? But what if there was such a device?

The truth is, you can forget about a transporter. No one has been able to realize such a concept. But that doesn't mean some of the ideas that seemed far-fetched when the show debuted in 1966 haven't become a reality. What technologies have actually come to pass?

In this section, we feature the top ten technologies from *Star Trek* that actually did come to fruition, listed in no particular order. Some of them may surprise you!

1. Transparent Aluminum (Armor)

The fourth installment of the original *Star Trek* movies is perhaps the most endearing to fans. The crew returns to modern-day Earth. Kirk, Spock, and the rest of the gang ditch a Klingon Bird of Prey spacecraft in the San Francisco Bay after narrowly missing the Golden Gate Bridge while flying blind in a storm. You may remember the scene—but do you remember Scotty introducing transparent aluminum for the first time?

In the flick, Scotty traded the formula matrix for transparent aluminum—a huge engineering advancement—for sheets of Plexiglas in order to build a tank to transport the two humpback whales (George and Gracie) to the Earth of their time. The claim was that you'd be able to replace six-inch (14-centimeter) thick Plexiglas with one-inch (2.5-centimeter) thick see-through aluminum.

It may sound impossible, but there is such a thing as transparent aluminum armor, or aluminum oxynitride (ALON) as it's more commonly known. ALON is a ceramic material that starts out as a powder before heat and pressure turn it into a crystalline form similar to glass. Once in the crystalline form, the material is strong enough to withstand bullets. Polishing the molded ALON strengthens the material even more. The Air Force has tested the material in hopes of replacing windows and canopies in its aircraft. Transparent aluminum armor is lighter and stronger than bulletproof glass. Less weight, stronger material—what's not to like?

⚡2. Communicators

Whenever Captain Kirk left the safe confines of the *Enterprise*, he did so knowing it could be the last time he saw his ship. Danger was never far away. And when in distress and in need of help in a pinch, he could always count on Bones to come up with a miracle cure, Scotty to beam him up, or Spock to give him some vital scientific information. He'd just whip out his communicator and place a call.

Fast forward thirty years, and—wouldn't you know it—nearly everyone carries a communicator. We just know them as cell phones. Actually, the communicators in *Star Trek* were more like the push-to-talk, person-to-person devices first made popular by Nextel in the mid- to late '90s. The *Star Trek* communicator had a flip antenna that, when opened, activated the device. The original flip cell phones are perhaps distant cousins. Whatever the case, the creators of *Star Trek* were on to something because you'd be hard pressed to find many people without a cell phone these days.

BADGE COMMUNICATION

In later incarnations of the *Star Trek* franchise, the communicators evolved to being housed in the Starfleet logo on the crewman's chest. With the tap of a finder, communication between crewmembers became even easier. Vocera Communications has a similar product, one that can link people on the same network, inside a designated area like an office or a building, by using the included software over a wireless LAN. The B2000 communication badge weighs less than two ounces, can be worn on the lapel of a coat or shirt, and allows clear two-way communication. It's even designed to inhibit the growth of bacteria so it's suitable for doctors.

⇵3. Hypospray

The creative team behind *Star Trek* found spiffy ways to spice up some activities we endure on a day-to-day basis. Take medical treatment, for example. Not many people enjoy getting a flu shot, and on *Star Trek*, inoculating patients was one of Dr. Leonard "Bones" McCoy's primary duties. It seemed not an episode went by that Bones wasn't giving someone a shot of some sort of space vaccine. But what was more fascinating was the contraption he used.

Hypospray is a form of hypodermic injection of medication. A hypospray injection is forced under the skin (a subcutaneous injection) with high-pressure air. The air pressure shoots the liquid vaccine deep enough into the skin that no needle is required. The real-world application is known as a jet injector.

Jet injectors have been in use for many years. In fact, the technology predates *Star Trek*. Jet injectors were originally designed to be used in mass vaccinations. Jet injecting is safer (no needles to pass along infectious disease) and faster in administering vaccines. Similar in appearance to an automotive paint gun, jet injection systems can use a larger container for the vaccine, thus allowing medical personnel to inoculate more people more quickly. Pretty cool, huh?

⚡ 4. Tractor Beams

When NASA needs to make repairs to the Hubble Space Telescope, astronauts have to be specially trained to get out of the Space Shuttle for extra-vehicular activity. They also have to learn how to work within the confines of their space suits, with thick gloves on. Wouldn't it be nice to just bring the telescope inside, where repairs wouldn't be so challenging and dangerous?

In science fiction, spaceships, including the Starship *Enterprise*, snatch each other up using tractor beams. In some cases, large vessels have tractor beams strong enough to prevent smaller vessels from escaping their gravitational force. But is this science even plausible?

Yes and no. Optical tweezers are as close as you're

going to get to a legitimate tractor beam on current-day Earth. Scientists have harnessed small lasers into beams capable of manipulating molecules and moving them with precision. Optical tweezers use a focused laser to trap and suspend microscopic particles in an optical trap. Scientists can use optical tweezers to trap and remove bacteria and sort cells. Optical tweezers are used primarily in studying the physical properties of DNA. While the beams used in optical tweezers aren't strong enough to dock the space shuttle to the International Space Station, they're a start in that direction.

5. Phasers

"Set phasers to stun," was an frequently heard command given to the *Enterprise* crew. The crew relied heavily on the stun setting of their fictitious weapon of choice, known as a phaser. Armed with a phaser, Kirk and his colleagues had the ability to kill or, more desirably, stun their adversaries and render them incapacitated.

Actually, stun guns have been around for some time. In fact, electricity was used for punishment and to control livestock as far back as the 1880s. But it wasn't until 1969, when a guy named Jack Cover invented the first Taser, that the stun gun was most realized. The Taser is not designed to kill, like the phaser did. However, it packs enough of an electrical punch to make its victim disorientated, if not completely incapacitated.

WITHIN ARM'S LENGTH

Unlike the phaser, the Taser and other stun guns must come in physical contact with the target to have any effect. Tasers take care of this by projecting two electrodes, connected to the gun by wires, that attach to the target's skin. Once contact has been made, the handheld unit transfers electricity to the target, thus creating the stun effect. Stun guns with stationary electrical contact probes are somewhat less effective than phasers because, while they have a similar effect on the target, you have to be much closer (within arm's length) in order to zap your target.

Something more along the lines of the phaser may be in development. Applied Energetic has developed laser-guided energy and laser-induced plasma energy technologies that are said to transmit high-voltage bursts of energy to a target. In other words, these pulses of energy would stun the target and limit collateral damage. So a true phaser may soon be a reality.

⇌ 6. Universal Translator

Imagine if, no matter what country you visited, no matter what the culture, you could understand everything the indigenous people were saying. It sure would make traveling easier. Take that thought to another level, perhaps if you were planet-hopping like the crew of the *Enterprise*. Fortunately for Captain Kirk and his peers, they had a universal translator.

The characters in *Star Trek* relied on a small device that, when spoken into, would translate the words into English. Guess what? The technology exists for us in the real world. There are devices that let you speak phrases in English and then spit back to you the same rhetoric in a specified language. The problem is that these devices only work for certain predetermined languages.

A true universal translator like the one on the show may not be a reality, but the technology is available. Voice recognition has advanced considerably since its inception. And smartphones and websites like BabelFish and Google are making it easier to translate entire articles, websites, and even books in an instant.

But these technologies—and computers in general—have yet to be able to learn *all* languages and translate accurately between any two given tongues. Computers would theoretically be able to gather the linguistic information much faster than a human brain, but a software program is dependent on actual data. Someone has to take the time and cover the expense to put it together and make it available, which is probably why these systems focus on the most popular languages.

⇟7. Geordi's VISOR

When *Star Trek: The Next Generation* thrust the love of everything *Star Trek* back into popular culture, the quirky Mr. Spock and crass Bones McCoy and others were supplanted by a new cast. One of the most popular characters on the new show was engineer Geordi LaForge.

What made Geordi unique, perhaps even mysterious, was his funky eyewear. Geordi was blind, but after a surgical operation and aided through the use of a device called VISOR (Visual Instrument and Sensory Organ Replacement), Geordi could see throughout the electromagnetic spectrum. Though it may sound far-fetched, in reality, similar technology exists that may someday bring sight back to the blind.

In 2005, a team of scientists from Stanford University successfully implanted a small chip behind the retina of blind rats that enabled them to pass a vision recognition test. The science behind the implants, or bionic eyes as they're commonly referred to, works much the way Geordi's VISOR did.

The patient receives the implants behind the retina and then wears a pair of glasses fitted with a video camera. Light enters the camera and is processed through a small wireless computer, which then broadcasts it as infrared LED images on the inside of the glasses. Those images are reflected back into the retina chips to stimulate photodiodes. The photodiodes replicate the lost retinal cells, then change light into electrical signals which, in turn, send nerve pulses to the brain.

What it all means is that in theory, a person with 20/400 sight (blind), due to the loss of retinal cells from retinitis pigmentosa, can obtain 20/80 sight. It's not good enough to pass the driving test (normal vision is considered 20/20), but it's good enough to read billboards and go about your day without the aid of a seeing-eye dog.

⇟ 8. Torpedo Coffins

In the second installment of the *Star Trek* movie franchise, the beloved Mr. Spock, played by Leonard Nimoy, died after saving the Starship *Enterprise* from certain disaster. The movie culminated with the crew firing Spock's corpse out of the torpedo bay in a coffin shaped like one of the ship's weapons, the photon torpedo.

Believe it or not, you too could be laid to eternal rest in your own Federation-approved photon torpedo casket.

Okay, it may not technically be Federation-approved since there is no such thing as the United Federation of Planets, but the coffins are, in fact, very real.

Designed by Eternal Image, the Star Trek coffin was slated to be available early 2009 but is still not for sale as of this writing. The price is yet to be determined. If the fan would prefer to be cremated, the company also plans to offer a *Star Trek* urn.

9. Telepresence

In 1966, the idea of interacting with each other while separated by the void of space seemed as far-fetched as the idea of being in space. That's precisely what the idea of telepresence is.

Telepresence is more than just videoconferencing. The visual aspect is important and immersion is vital. In other words, the more convincing the illusion of telepresence, the more you feel like you're there.

In 2008, AT&T teamed up with Cisco to deliver the industry's first in-depth telepresence experience. The key to Cisco's TelePresence is the combination of audio, video, and ambient lighting working together. These telepresence kits are designed to mirror surroundings and mimic sounds so that users on each side of the videoconference will feel as though the images on the screen are in the same room with them.

For instance, the people in boardroom A will see the people on the screen in boardroom B as though they were sitting across the table from them. The ambient lighting and room features are constructed to mirror each other. Sure, these telepresence kits are much more advanced than anything drummed up on *Star Trek*, but perhaps that's because the show sparked our imagination so many years ago.

⇟ 10. Tricorders

How many of you remember that instrument Mr. Spock used to always carry over his shoulder, especially when the crew (usually consisting of only Spock and Captain Kirk) first surveyed a new planet? That was a tricorder.

One of the more useful instruments available to Star Trek personnel, variations of the tricorder (medical, engineering, or scientific) were used to measure everything from oxygen levels to detecting diseases. Often the tricorder gave an initial analysis of the new environment. So, what's the real-world tie-in? NASA employs a handheld device called LOCAD, which measures for unwanted microorganisms such as *E. coli*, fungi, and salmonella onboard the International Space Station. Beyond that, two handheld medical devices may soon help doctors examine blood flow and check for cancer, diabetes, or bacterial infection.

Scientists at Loughborough University in England use

photoplethysmography technology in a handheld device that can monitor the functions of the heart. Meanwhile, researchers at Harvard Medical School have developed a small device that uses similar technology to that found in MRI machines in order to non-invasively inspect the body. Using nuclear magnetic resonance imaging, this device would be sensitive enough to measure samples of as few as ten possible infectious bacteria. This kind of sensitivity (eight hundred times more sensitive than sensing equipment currently used in medical labs) could revolutionize the way doctors diagnose disease.

All of the technological advancements in this section prove that when we humans put our minds to it, there are lots of super-technologies we can create to give ourselves various super-abilities. In Part II, we'll look at more scientific advances that are getting us closer and closer to becoming real superhumans.

QUIZ: THE SUPER SCI-FI HEROES AND VILLAINS QUIZ

So you've learned a bit of the backstories about some of your favorite superheroes and supervillains so far. But what if you had to face them on the street? Once you pass this quiz, you'll be prepped for any questions they might throw your way. (Though we can't guarantee you'll be prepped for any superpowers they throw at you!) Check your answers in the back of the book on page 281.

1. Let's start with a classic backstory—the orphaned superhero. Which superhero orphan was taken in by a nice couple named Jonathan and Martha?

   a. Batman

   b. Superman

   c. Yoda

2. What color was not used as a color for kryptonite in any version of the Superman series?

   a. Red

    b.   Green

    c.   Black

3. **According to the comparison done by HowStuffWorks, who of the following would stand a chance of winning against Superman in a battle of wills?**

    a.   Chuck Norris

    b.   A team of the X-Men

    c.   A Sith Lord

    d.   All of the above

4. **Which of the below is NOT part of the over-armor of Batman's awesome Batsuit?**

    a.   Calf guards

    b.   Ankle guards

    c.   A spine guard

5. **Which trait is NOT shared by Batman and Superman?**

    a.   Being orphans

    b.   Having a secret identity

    c.   Having superpowers

6. **What is the sonic device that Batman has?**

   a. A device that can be used to summon bats instantly.

   b. A device that can be used to call for help.

   c. A device that can be used to contact Commissioner Gordon directly.

7. **True or False: The Batmobile has a jet engine specifically designed to let it blast through walls or any other objects in its way.**

   a. True

   b. False

8. **In the most accepted version of the Joker's origins, how does the Joker turn into the Joker?**

   a. He was violently attacked by a clown as a boy and became obsessed with exacting murderous revenge on society after that.

   b. He was a failed comedian who went on a murderous rampage when his audiences refused to give him the encouragement and fame he sought.

   c. He was transformed into a crazy man after he jumped into a vat of chemicals during a botched robbery at a Gotham chemical plant.

9. **What is the major weapon on the superweapon, the Death Star, that the Sith created?**

   a.   The superlaser

   b.   The tractor beam system

   c.   The ion cannons

10. **True or False: Einstein's special theory of relativity makes it possible for the Starship *Enterprise* to use its superability, warp speed, to travel faster than the speed of light.**

   a.   True

   b.   False

BECOMING SUPERHUMAN:
HOW WE CAN GAIN OUR OWN SUPERPOWERS

So far we've heard about teleporting, flying, traveling at warp speed, and using lots of super-cool weapons to defeat supervillains. A lot of these superhero stories have to do with war, and that makes sense. As a species, humans are almost constantly at war. But what about making our own bodies superhuman so we can protect them against the casualties of war and combat?

Could we ever regrow limbs or create our own human body armor that conforms to our skin? And what about our creations that seem human but aren't, things like replicants and Cylons? Are they possibilities in the future, and if so, would they be friends or enemies? In this part, we'll take a look at what the future may hold for us and our robotic counterparts in the battle to become superhuman.

Let's start by taking a cue from Mother Nature herself. In the first section, we'll check out the lowly salamander, which has a remarkable skill: it can regrow parts of its own body. Is this something we could do, too, one day?

NO MORE PHANTOM LIMBS: HOW CAN SALAMANDERS REGROW BODY PARTS?

American military armor is better designed than ever before to protect soldiers on the battlefield from being killed. While that has drastically lowered the number of U.S. soldier casualties in the wars in Iraq and Afghanistan, thousands of soldiers are returning stateside with serious burns, missing limbs, and other debilitating injuries.

To improve treatment options for veterans, the Pentagon recently announced its plan to devote $250 million to research on regenerating human skin, ears, and muscles for injured soldiers. Part of that funding will also establish the Armed Forces Institute of Regenerative Medicine that will focus on human limb regeneration, among other things.

LESSONS FROM A LIZARD

If a salamander gets in a scrape, it can drop its tail, scurry off, and return to business as usual. What if we could do that? Scientists are using the salamander as a blueprint for human genome research to reveal how to regenerate human limbs.

Limb regeneration doesn't mean growing arms and legs in test tubes. Instead, it means that a person would actually regrow a limb themselves. Scientific evidence indicates that humans have the potential for limb regeneration in our genes as salamanders do, but those genes are dormant in our bodies. Human embryos, for instance, can regrow limb buds in the womb. And a man in Cincinnati, Ohio, regrew a fingertip after accidentally slicing it off in 2005. But when you lose an entire limb, the body reacts by covering that wound site with thick scar tissue to ward against infection.

To figure out how we might be able to reignite that genetic potential for limb regeneration, researchers are starting small—with mice. But they don't have to work entirely from scratch to track down how an organism could regrow something. They're looking to the salamander as their model. How does this relatively simple creature perform a science fiction-style anatomical magic trick?

⚡Salamander Limb Regeneration

If a salamander gets in a fight, it may surrender its tail to the enemy as a defense mechanism. After all, in a few weeks' time, it can grow a new one. This is a fairly complex process, but in a nutshell, regeneration involves shuffling around the cells at the wound site and assigning them a new specialization.

SALAMANDER SPECIFICATIONS

Salamanders are part of the amphibian family, members of which are cold-blooded and have an additional skin covering of feathers or fur. Different species of salamanders are either terrestrial or aquatic and are the only amphibians with tails. In case they lose that precious tail, salamanders can grow it back. They're the highest order of animals capable of regenerating body parts, including their tails, upper and lower jaws, eyes, and hearts.

Within the first hours after a body part is lopped off, the salamander's epidermal cells in the area migrate to cover the open flesh. That layer of cells gradually thickens in the following days, forming the apical epithelial cap. Cells called fibroblasts found within the salamander's tissues also congregate beneath that epidermal covering. Fibroblasts are undifferentiated, which means that they're free to become multiple types of cells, depending on which body part needs replacing.

After that initial phase, the blastema develops from the mass of fibroblasts. The blastema will eventually become the replacement body part. Researchers recently discovered that the expression of a protein called nAG kick-starts blastema growth. The blastema is sort of like a mass of human stem cells in that it has the potential to grow into various limbs, organs, and tissues. But how does the salamander's body know what needs replacing? The genetic coding in the blastema contains a positional memory about the location and type of missing body part. That data is stored in the Hox genes in the fibroblast cells.

While this is happening, capillaries and blood vessels are regenerating into the blastema. As the blastema cells divide and multiply, the resulting mass becomes a bud of undifferentiated cells. In order for that growth to become a full-fledged limb, tail, or other body part, it must receive stimulation from nerves. However, when salamanders drop their tails, they lose not only flesh but also nerves. That means that nerve axon regeneration is happening at the wound site simultaneously with tissue, bone, and muscle regeneration.

From there, the cells differentiate and create the

appropriate body part. As part of that positional memory in the fibroblast cells, the blastema knows to grow in the proper sequence to avoid defective regeneration. For example, if a salamander loses a foot at its ankle, the blastema will develop outward to form a foot instead of an entire leg.

With the salamander as the blueprint, scientists hope to someday engineer blastemas from human cells, which would help amputees regrow limbs and heal from injuries. Until then, our amphibian friends are still the reigning regenerators of the animal kingdom.

In the meantime, however, we are making other significant improvements in body armor that can not only help soldiers and law enforcement officials avoid amputations and limb loss in combat or on the job, but also give them abilities and skills beyond those of any ordinary human. Let's take a look at how these force warriors will look in the future!

HOW THE FUTURE FORCE WARRIOR WILL WORK

As we've seen in both real life and in science fiction, wars are evolutionary, with each new conflict bringing more powerful and advanced weaponry. Weapons that yield success on the modern battlefield can become outdated and ineffective in just a few years. The reality of the battlefield necessitates continuous change to stay a step ahead of the enemy.

To better equip its soldiers, the U.S. Army is developing an advanced infantry uniform that will provide superhuman strength and greater ballistic protection than any uniform to date. Also, using wide-area networking and onboard computers, soldiers will be more aware of the action around them and of their own bodies.

In the next few sections, we will leap from regenerating limbs onto the battlefield of the future to see how its technology compares with today's and learn how the future force warrior will turn a soldier into an "F-16 on legs." First, let's take a look at the most fundamental identity factor of a soldier: the uniform.

⇌ Suiting Up for Battle

With the development of a bionic uniform for its soldiers, the U.S. Army is planning for a change in the logistics of war. Integrated physiological monitoring, enhanced communication, and augmented physical strength will give the soldiers of the future the tools they need to overwhelm their opponents simply by donning a high-tech suit.

TWO-PHASE PROGRAM

There are two phases to the Future Force Warrior program. The first phase involved the deployment of a uniform in 2010 which met the Army's short-term needs, although pieces of the uniform may be deployed earlier. According to Future Force Warrior equipment specialist Jean-Louis "Dutch" DeGay, "The Department of the Army has built what are called design spirals, so roughly every two years, if a piece of technology has matured, we try to get it in the field, rather than waiting to field the entire system." In 2020, the U.S. Army will roll out a suit that integrates nanotechnology, exoskeletons, and liquid body armor, all of which exist only in concept now.

Here are the basic components of the final version of this superhuman suit:

- **Helmet**—The helmet houses a GPS receiver, radio, and the wide- and local-area network connections.
- **Warrior Physiological Status Monitoring System**—This layer of the suit is closest to the body and contains sensors that monitor physiological indicators, such as heart rate, blood pressure, and hydration. The suit relays the information to medics and field commanders.
- **Liquid Body Armor**—This liquid body armor is made from magnetorheological fluid, a fluid that remains in a liquid state until the application of a magnetic field. When an electrical pulse is applied, the armor transitions from a soft state to a rigid state in thousandths of a second.
- **Exoskeleton**—The exoskeleton is made of lightweight, composite devices that attach to the legs and augment the soldier's strength.

Together, these subsystems combine to create a uniform that informs, protects, and enhances the abilities of its wearer. Now let's take of each of these components separately.

Battlefield Awareness

The value of reconnaissance depends on how quickly that

information can be relayed to the soldier on the battlefield. The soldiers of the future will have more information immediately available to them than ever before.

The U.S. Army currently employs a system called Blue Force Tracker (BFT). The system enables a commander to get a real-time picture of the battlefield from his or her personal computer. The commander can then track individual unit movement and provide this information to friendly units. The U.S. Marines have used BFT, although they initially opted for a more portable and rugged system called the Enhanced Position Location Reporting System, or "ePLRS." Both ePLRS and BFT share the same goal: real-time tracking of friendly forces. The downside to both systems, however, is that they are bulky, somewhat dated, and require computers with operators who could otherwise be carrying a weapon.

The Future Force Warrior setup is a significant improvement on these current systems. A computer embedded in the suit and located at the base of the soldier's back will be connected to a local and wide-area network, allowing for data transfer. And in fact, it derives some inspiration from *Star Trek*, as "Dutch" DeGay explains:

> *Essentially, it's what we call the "borg" effect, to borrow a theme from Star Trek. Everything in the battle space is a sensor, whether that's a vehicle, rotor*

wing, fixed wing, aviation vehicle, ground vehicle,
individual soldier, or unmanned robotic platform. That
becomes a sensor that I can track for data. I can send
data to it or take data, video, or audio from it.

Soldiers will use a voice-activated, drop-down screen in
the helmet to access information without having to put down
their weapons. Embedded in a pair of transparent glasses, the
display will appear to the soldier as a 17-inch screen. This
screen can display maps and real-time video provided by a
forward-positioned scout team, satellite, or aircraft.

According to DeGay, "We are working to have the
graphic user interface inside the computer systems to either
replicate computer graphic user interfaces or even Playstation
2 or Xbox graphic user interfaces," because most of today's
soldiers are already familiar with how those systems work.

YOU OK? YES, YOU ARE.

Not only will Future Force Warriors know more about
their fellow soldiers, but they also will know more about
their own physiological condition. The physiological
subsystem of the uniform lies against the soldier's
skin and includes sensors that monitor soldier's core
body temperature, skin temperature, heart rate, body

position (standing or sitting), and hydration levels. These statistics are monitored by the soldier and by medics and commanding officers who might be miles away. Knowing the condition of a platoon of soldiers allows commanders to make better strategic decisions. The Future Force Warrior helmet also includes a GPS receiver, providing commanders with exact positioning data on their troops.

Another vital component of battle is communication among soldiers. The Future Force Warrior will use sensors that measure vibrations of the cranial cavity, eliminating the need for an external microphone. This bone-conduction technology allows soldiers to communicate with one another, and it also controls the menus visible through the drop-down eyepiece. The helmet has 360-degree situational awareness and voice amplification.

"What this will allow you to do is to know where that sniper round or mortar round came from, but at the same time it will cancel out noise at a certain decibel so as to not cause damage to the soldier's ears," said Robert Atkinson, liaison sergeant with the Operational Forces Interface Group at Natick Soldier Center.

The situation-awareness technology also allows soldiers to:

* Detect other soldiers in front of them up to a couple of kilometers away.
* Focus in on a particular sound and amplify it.

Powering the entire suit is a 2- to 20-watt microturbine generator fueled by a liquid hydrocarbon. A plug-in cartridge containing 10 ounces of fuel can power the soldier's uniform for up to six days. Battery patches embedded in the helmet provide three hours of backup power.

Liquid Body Armor

We'll explore this mind-blowing technology in a later section, but, long story short, with advances in ballistics, armies must develop better body armor, far beyond the Kevlar we're used to. Scientists are now working on a new breed of armor made from magnetorheological (MR) fluids—*liquid body armor*.

One type of MR fluid consists of small iron particles suspended in silicon oil. The oil prevents the particles from rusting. The fluid transforms from liquid to solid in just milliseconds when a magnetic field or electrical current is applied to it. The current causes the iron particles to lock into a uniform polarity and stack on top of each other, creating an impenetrable shield. How hard the substance becomes depends on the strength of the magnetic field or electrical current. Once the charge or magnetic field is removed, the particles unlock, and the substance goes back to a fluid state.

MR fluid will fill small pockets in the Future Force Warrior uniform fabric. The uniforms will be wired to allow an electrical current to pass through the fabric. The electrical current will be controlled by the onboard computer system and will automatically charge the MR fluid when there is a ballistic threat present.

Scientists at Massachusetts Institute of Technology who are developing the liquid body armor say that it will take five to ten years to make the substance fully bullet resistant.

⚡Exoskeleton

Superhuman strength has always been confined to science fiction, but advances in human-performance augmentation systems could give soldiers the ability to lift hundreds of pounds using the effort they would usually use to lift a fraction of that weight.

The shoulder of the Future Force Warrior uniform contains fabric filled with nanomachines that mimic the action of human muscles, flexing open and shut when stimulated by an electrical pulse. These nanomachines will create lift the way muscles do and augment overall lifting ability by 25 to 35 percent.

"Think of yourself on steroids, holding as much weight as you want for as long as you want," said Atkinson. "It will also allow a 90-pound male or female to carry a 250-pound male or female off of the battlefield and it wouldn't feel like they were carrying 250 pounds worth of person."

LIFTING AND LOAD CARRYING x 3

The exoskeleton attached to the lower body of the soldier will provide even more strength. The overall exoskeleton will provide up to 300 percent greater lifting and load-carrying capability.

"The exoskeleton, which is [being developed] in conjunction with DARPA, will give the soldier more stability," Atkinson said. "It makes the soldier become a weapons platform." (DARPA is the U.S. Department of Defense's Defense Advanced Research Projects Agency.)

With this added strength, weapons can be mounted directly to the uniform system.

The exoskeleton will merge structure, power, control, actuation, and biomechanics. Here's a look at some of the challenges that DARPA has outlined:

- **Structural materials**—The exoskeleton will have to be made out of composite materials that are strong, lightweight, and flexible.

- **Power source**—The exoskeleton must have enough power to run for at least twenty-four hours before refueling.

- **Control**—Controls for the machine must be seamless. Users must be able to function normally while wearing the device.

- **Actuation**—The machine must be able to move smoothly so it's not too awkward for the wearer. Actuators must be quiet and efficient.

- **Biomechanics**—Exoskeletons must be able to shift from side to side and front to back, just as a person would move in battle. Developers will have to design the frame with human-like joints.

As warfare changes, armies are looking for any advantage they can get against potential enemies. The new Future Force Warrior suit will take human performance to unprecedented levels. Imagine a platoon of soldiers wearing suits that turn an ordinary person into a real, live superhero.

Thank you to 1st Lieutenant John H. Frushour, USMC, 24th Marine Expeditionary Unit, for his assistance with this section.

I'M INVINCIBLE! HOW BODY ARMOR WORKS

Humans have been wearing armor for thousands of years. Ancient tribes fastened animal hide and plant material around their bodies when they went out on the hunt, and the warriors of ancient Rome and medieval Europe covered their torsos in metal plates before going into battle. By the 1400s, armor in the Western world had become highly sophisticated. With the right armor, you were nearly invincible.

All that changed with the development of cannons and guns in the 1500s. These weapons hurled projectiles at a high rate of speed, giving them enough energy to penetrate thin layers of metal. You can increase the thickness of traditional armor materials, but they soon become too cumbersome and heavy for a person to wear.

It wasn't until the 1960s that engineers developed reliable, bullet-resistant armor that a person could wear comfortably. Unlike traditional armor, this soft body armor is not made out of pieces of metal; it is formed from advanced woven fibers that can be sewn into vests and other soft clothing. But how on earth can something soft like that protect you from a piercing bullet?

CERAMICS?

Why would body armor be made with ceramic plates? Bathroom tile is made of ceramic, and it is extremely brittle and fragile. What good is it in body armor?

Thousands of different materials are classified as ceramics. The ceramic used in body armor is called alumina, with the chemical formula Al_2O_3. Sapphires are made of alumina, and sapphire is a very hard material.

You can also find rigid plates made out of the plastic polyethylene. It is thicker than ceramic and not quite as strong, but lighter.

⚡Soft Can Be Hard: How Soft Body Armor Works

Soft body armor is a fairly mystifying concept: how can a soft piece of clothing stop bullets? The principle at work is actually quite simple. At its heart, a piece of bulletproof material is just a very strong net.

To see how this works, think of a soccer goal. The back of the goal consists of a net formed by many long lengths of tether, interlaced with each other and fastened to the goal frame. When you kick the soccer ball into the goal, the ball has a certain amount of energy, in the form of forward inertia. When the ball hits the net, it pushes back on the tether lines at that particular point. Each tether extends from one side of the frame to the other, dispersing the energy from the point of impact over a wide area.

The energy is further dispersed because the tethers are interlaced. When the ball pushes on a horizontal length of tether, that tether pulls on every interlaced vertical tether. These tethers in turn pull on all the connected horizontal tethers. In this way, the whole net works to absorb the ball's inertial energy, no matter where the ball hits.

If you were to put a piece of bulletproof material under a powerful microscope, you would see a similar structure. Long strands of fiber are interlaced to form a dense net. A bullet is traveling much faster than a soccer ball, of course, so the net needs to be made from stronger material. The most famous material used in body armor is DuPont's Kevlar fiber. Kevlar is lightweight, like a traditional clothing fiber, but it is five times stronger than a piece of steel of the same weight. When interwoven into a dense net, this material can absorb a great amount of energy.

In addition to stopping the bullet from reaching your body, a piece of body armor also has to protect against blunt trauma caused by the force of the bullet. In the next section, we'll see how soft body armor deals with this energy so that the wearer doesn't suffer severe injuries.

BEYOND KEVLAR

Kevlar is by far the most common fiber used to make body armor, but other materials are being developed.

The most readily available alternative fiber is called Vectran, which is approximately twice as strong as Kevlar. Vectran is five to ten times stronger than steel.

Another rapidly emerging fiber is spider silk. Yes, spider silk. Goats have been genetically engineered to produce the chemical constituents of spider silk, and the resulting material is called Biosteel. A strand of Biosteel can be up to twenty times stronger than an equivalent strand of steel. Chicken feathers are also a possibility. University of Nebraska-Lincoln researchers are spinning them into cloth that is lightweight and very sturdy. Because the feathers have a fine honeycomb texture, they could be resistant to bullets.

Another candidate is carbon nanotubes, which promise to be even stronger than spider silk. Carbon nanotube thread is still rare, and fabric made from it is even rarer. CNET reports the current price of nanotubes at $500 per gram. In time, prices should fall and make carbon nanotubes a viable fiber for body armor.

⚡Raining Blows Down on You: Blunt Trauma and Ranking Resistance

In the last section, we saw that a piece of soft bulletproof material works in the same basic way as the net in a soccer goal. Like a soccer goal, it has to "give" a certain amount to absorb the energy of a projectile.

When you kick a ball into a soccer goal, the net is pushed back fairly far, slowing the ball down gradually. This is a very efficient design for a goal because it keeps the ball from bouncing out into the field. But bulletproof material can't give as much because the vest would push too far into the wearer's body at the point of impact. Focusing the blunt trauma of the impact in a small area can cause severe internal injuries.

Bulletproof vests have to spread the blunt trauma out over the whole vest so that the force isn't felt too intensely in any one spot. To do this, the bulletproof material must have a very tight weave. Typically, the individual fibers are twisted, increasing their density and their thickness at each point. To make it even more rigid, the material is coated with a resin substance and sandwiched between two layers of plastic film.

A person wearing body armor will still feel the energy of a bullet's impact, of course, but over the whole torso rather than in a specific area. If everything works correctly, the victim won't be seriously hurt.

Since no one layer can move a good distance, the vest has to slow the bullet down using many different layers. Each "net" slows the bullet a little bit more until the bullet finally stops. The material also causes the bullet to deform at the point of impact. Essentially, the bullet spreads out at the tip, in the same way a piece of clay spreads out if you throw it against a wall. This process, which further reduces the energy of the bullet, is called "mushrooming."

No bulletproof vest is completely impenetrable, and there is no piece of body armor that will make you invulnerable to attack. There's actually a wide range of body armor available today, and the types vary considerably in effectiveness.

CATEGORIZING BODY ARMOR

In the United States, body armor levels are certified by the National Institute of Justice (NIJ), which is an agency of the U.S. Department of Justice. The levels are I, II-A, II, III-A, III, and IV. Based on extensive laboratory tests, researchers classify any new body-armor design into one of seven categories, with Category I body armor offering the lowest level of protection and Category IV the highest.

The body-armor classes are often described by the sort of weaponry they guard against. The lowest-level body armor can only be relied on to protect against bullets with a relatively small caliber (diameter), which tend to have less force on impact. Some higher-grade body armor can protect against powerful shotgun fire. Categories I through III-A are soft and concealable. Category III is the first one to use hard or semi-rigid plates.

Generally speaking, armor with more layers of bullet-proof material offers greater protection. With some bullet-proof vests, you can add layers. One common design is to fashion pockets on the inside or outside of the vest. When you need extra protection, you insert metal or ceramic plates into the pockets. When you don't need as much protection, you can wear the vest as ordinary soft armor.

To determine how effective a particular armor design is, researchers shoot it with all sorts of bullets, at all sorts of angles and distances. For a piece of armor to be considered effective against a particular weapon at a particular range, it has to stop the bullet without causing dangerous blunt trauma. The researchers determine blunt trauma by molding a layer of clay onto the inside of the armor. If the clay is deformed more than a certain amount at the point of impact, the armor is considered ineffective against that weaponry.

⚡Your Armor of Choice

It may seem odd that a police officer would wear Category I body armor, which will only stop relatively small-caliber bullets, when they could have superior protection from higher-ranked armor. But there is a very good reason for this decision. Typically, higher-ranked armor is a lot bulkier and heavier than lower-ranked armor, which results in several problems:

* An officer has reduced flexibility and comfort in bulkier armor, which impedes police work. You can't chase a criminal very well when you're carrying a massive weight on your torso.

* Heavier armor may actually increase the chances of an officer being severely wounded. An attacker would be more aware of a heavy armored jacket than a thin concealed vest and, therefore, might aim at an unarmored part of the body, such as the head.

* The discomfort of heavier armor makes it more likely

that an officer won't wear any protection at all. Police departments are very careful to select bulletproof vests that are relatively comfortable to encourage officers to actually put them on.

Armor effectiveness and comfort are sure to improve in the future as technology companies develop lighter, stronger materials. We are certainly a long way from impenetrable armor; but in fifty years, advanced armor will give police officers a much greater level of protection when they're walking the beat.

Most likely, we will also see an increase in civilian body armor in the years ahead. There is an ever-growing market for comfortable soft-body armor that can fit under clothes or even be worn as an outer jacket. With gun violence on the rise, many citizens feel as if they're walking onto a battlefield every day, and they want to dress accordingly.

FLUID FORTIFICATION: HOW LIQUID BODY ARMOR WORKS

As we saw in the previous section, the basic idea behind body armor hasn't changed much in the past few thousand years.

- First, armor stops weapons or projectiles from reaching a person's body.
- Second, it diffuses the weapon's energy so that the final impact causes less damage.

While it's not effective in every situation, armor can generally help protect people from serious injury or death, especially against the right weaponry. But over the years, people have had to develop stronger and more advanced armor to protect against increasingly sophisticated weapons.

However, in spite of these improvements, modern body armor still has some of the same shortcomings as ancient forms of armor. Whether it's made from metal plates or layers of fabric, armor is often heavy and bulky. Many types are rigid, so they're impractical for use on arms, legs, and necks. For this reason, medieval suits of plate armor had gaps and joints to allow people to move around, and the body armor used today often protects only the head and torso.

As we learned earlier, one of the newest types of body armor, though, is both flexible *and* lightweight. Oddly enough, this improvement comes from the addition of liquid to existing armor materials. While it's not entirely ready for combat, laboratory research suggests that liquid body armor has the potential to be a good replacement for or supplement to bulkier vests. Eventually, soldiers, police officers, and others may be able to use it to protect their arms and legs. But how does this concept even work?

⚡Kevlar Comes to the Rescue...Again!

The two primary types of liquid body armor currently in development both start with a foundation of DuPont Kevlar, commonly used in bulletproof vests, as we learned earlier. When a bullet or a piece of shrapnel hits a Kevlar vest, the layers of material spread the impact over a large surface area. The bullet also stretches the Kevlar fibers, expending energy and slowing down in the process. The concept is similar to what happens when a car air bag spreads the impact and slows the movement of a person's torso during a collision.

Although Kevlar is a fabric, Kevlar armor does not move or drape the way clothing does. It takes between twenty and forty layers of Kevlar to stop a bullet, and this stack of layers is relatively stiff. It's also heavy—a vest alone often weighs more than 10 pounds (4.5 kilograms), even without ceramic inserts for additional protection.

Two different fluids, however, can allow Kevlar armor to use far fewer layers, making it lighter and more flexible. They have one thing in common—they react strongly in response to a stimulus. Next, we'll look at what these liquids are made of and why they react the way they do.

Fluid Movements: Shear-Thickening Fluid

The term "liquid body armor" can be a little misleading. For some people, it brings to mind the idea of moving fluid sandwiched between two layers of solid material. However, both types of liquid armor in development work without a visible liquid layer. Instead, they use Kevlar that has been soaked in one of two fluids.

- The first is a shear-thickening fluid (STF), which behaves like a solid when it encounters mechanical stress or shear. In other words, it moves like a liquid until an object strikes or agitates it forcefully. Then, it hardens in a few milliseconds.

- This is the opposite of a shear-thinning fluid, like paint, which becomes thinner when it is agitated or shaken.

LIKE CORNSTARCH AND WATER

You can see what shear-thickening fluid looks like by examining a solution of nearly equal parts of cornstarch and water. If you stir it slowly, the substance moves like a liquid. But if you hit it, its surface abruptly solidifies. You can also shape it into a ball, but when you stop applying pressure, the ball falls apart.

Here's how the process works. The fluid is a colloid, made of tiny particles suspended in a liquid. The particles repel each other slightly, so they float easily throughout the liquid without clumping together or settling to the bottom. But the energy of a sudden impact overwhelms the repulsive forces between the particles—they stick together, forming masses called hydroclusters. When the energy from the impact dissipates, the particles begin to repel one another again. The hydroclusters fall apart, and the apparently solid substance reverts to a liquid.

The fluid used in body armor is made of silica particles suspended in polyethylene glycol. Silica is a component of sand and quartz, and polyethylene glycol is a polymer commonly used in laxatives and lubricants. The silica particles are only a few nanometers in diameter, so many reports describe this fluid as a form of nanotechnology.

To make liquid body armor using STF, researchers first dilute the fluid in ethanol. They saturate the Kevlar with the diluted fluid and place it in an oven to evaporate the ethanol. The STF then permeates the Kevlar, and the Kevlar strands hold the particle-filled fluid in place. When an object strikes or stabs the Kevlar, the fluid immediately hardens, making the Kevlar stronger. The hardening process happens in mere milliseconds, and the armor becomes flexible again afterward.

In laboratory tests, STF-treated Kevlar is as flexible as plain, or neat, Kevlar. The difference is that it's stronger, so armor using STF requires fewer layers of material. Four layers of STF-treated Kevlar can dissipate the same amount of energy as fourteen layers of neat Kevlar. In addition, STF-treated fibers don't stretch as far on impact as ordinary fibers, meaning that bullets don't penetrate as deeply into the armor or a person's tissue underneath. The researchers theorize that this is because it takes more energy for the bullet to stretch the STF-treated fibers.

Research on STF-based liquid body armor is ongoing at the U.S. Army Research Laboratory and the University of Delaware. Researchers at Massachusetts Institute of Technology, on the other hand, are examining a different fluid for use in body armor. We'll look at their research next.

THE SLOW BLADE PENETRATES THE SHIELD

STF-based body armor has parallels in the world of science fiction. In the universe of Frank Herbert's *Dune*, a device called a Holtzman generator can produce a protective shield. Only objects moving at slow speeds may penetrate this shield. Similarly, slow-moving objects will sink through shear-thickening fluid without causing it to harden. In low-speed, or quasistatic, knife tests, a knife can penetrate both neat Kevlar and STF-treated Kevlar. However, the STF-treated Kevlar sustains slightly less damage, possibly because the fluid causes the fibers to stick together.

A Magnet for Bullets: How Magnetorheological Fluid Works

The other fluid that can reinforce Kevlar armor is magnetorheological (MR) fluid. MR fluids are oils that are filled with iron particles. Often, surfactants surround the particles to protect them and help keep them suspended within the fluid. Typically, the iron particles comprise between 20 and 40 percent of the fluid's volume.

The particles are tiny, measuring between 3 and 10 microns. However, they have a powerful effect on the fluid's consistency. When exposed to a magnetic field, the particles line up, thickening the fluid dramatically. The term "magnetorheological" comes from this effect. Rheology is a branch of mechanics that focuses on the relationship between force and the way a material changes shape. The force of magnetism can change both the shape and the viscosity of MR fluids.

The hardening process takes around twenty-thousandths of a second. The effect can vary dramatically depending on the composition of the fluid and the size, shape, and strength of the magnetic field. For example, MIT researchers started with spherical iron particles, which can slip past one another even in the presence of the magnetic field. This limits how hard the armor can become, so researchers are studying other particle shapes that may be more effective.

IRON FILINGS AND OIL

As with STF, you can see what MR fluids look like using ordinary items. Iron filings mixed with oil create a good representation. When no magnetic field is present, the fluid moves easily. But the influence of a magnet can cause the fluid to become thicker or to take a shape other than that of its container. Sometimes, the difference is very visually dramatic, with the fluid forming distinctive peaks, troughs, and other shapes. Artists have even used magnets and MR fluids or similar fer-rofluids to create works of art.

With the right combination of density, particle shape, and field strength, MR fluid can change from a liquid to a very thick solid. As with shear-thickening fluid, this change could dramatically increase the strength of a piece of armor. The trick is activating the fluid's change of state. Since magnets large enough to affect an entire suit would be heavy and impractical to carry around, researchers propose creating tiny circuits running throughout the armor.

Without current flowing through the wires, the armor would remain soft and flexible. But at the flip of the switch, electrons would begin to move through the circuits, creating a magnetic field in the process. This field would cause the armor to stiffen and harden instantly. Flipping the switch back to the off position would stop the current, and the armor would become flexible again.

In addition to making stronger, lighter, more flexible armor, fabrics treated with shear-thickening and magnetorheological fluids could have other uses. For example, such materials could create bomb blankets that are easy to fold and carry but can still protect bystanders from explosion and shrapnel.

Treated jump boots could harden on impact or when activated, protecting paratroopers' everyday boots. Prison guards' uniforms could make extensive use of liquid armor technology, especially since the weapons guards are most likely to encounter are blunt objects and homemade blades.

However, the technologies do have a few pros and cons. Here's a rundown:

* Neither type of armor is quite ready for battlefield use.
* STF-treated Kevlar armor is starting to be available.
* MR fluid may require another five to ten years of development before it can consistently stop bullets.

So we're still a little way away from seeing liquid body armor on the battlefield, but this once-fictional power is becoming more and more of a reality every day. Next, we'll look at the rise of a potential super foe that is also growing closer and closer to becoming a reality: the replicant.

OTHER USES FOR MR FLUIDS

MR fluids have numerous uses besides strengthening body armor. Their ability to change from liquids to semisolids almost instantly makes them useful for dampening impacts and vibrations in items like:

❋   Car shock absorbers.

❋   Washing machines.

❋   Prosthetic limbs.

❋   Bridges.

Since MR fluids can instantly and reversibly change shape, they could also be used to create scrolling Braille displays or reconfigurable molds.

INVASION OF THE HUMANOIDS: HOW REPLICANTS WORK

Los Angeles, 2019. You're walking down a city street. Dark skies overhead drip with acid rain. Monolithic buildings covered in neon advertising dominate the landscape. Even though many people have moved off-world to the colonies, the street is crowded.

Ahead you see a woman in a see-through plastic raincoat running toward you. You step out of the way and see that she's being followed by a man with a gun. He fires at her as she plows her way through one plate-glass window, then another, until she can run no longer. She lies on the concrete, surrounded by broken glass and blood.

Police turn over the already-stiff body and ask the man for his credentials. He's Rick Deckard, a police officer known as a blade runner. It was his job to track the woman down and "retire" her. But she's no human—it's a replicant, one of six who killed twenty-three people, jumped a shuttle, and came to Earth for unknown reasons.

Maybe moving off-world isn't such a bad idea after all.

This is the world of Sir Ridley Scott's *Blade Runner*. It's a dreary place, to be sure. People pack the streets tightly, and animals are all but extinct. Rain pours from the sky, and even when the sun is shining, it seems dark. Advertising screams, sometimes literally, from every direction. Flying cars take police officers—and few others—from place to place. It's a world of high technology and low empathy, not a very human place to live.

The escaped replicants have come here at their peril. They're illegal on Earth, under penalty of death. So why have they come here? What do they want?

In this section, we'll look at exactly what replicants are and why they run. Then, we will look at the world of *Blade Runner*, and the androids and humans who live there, and check out whether we could one day live alongside robots, peacefully or not.

⚡What's a Replicant?

Replicants are androids, or robots in human form. Created by Eldon Tyrell and the Tyrell Corporation in the *Blade Runner* future, they can do many different kinds of work. They're especially well-suited for hard labor or jobs that are too hazardous for humans to do. Advertisements for moving to the off-world colonies promote the opportunity to own a replicant as an incentive. There are also other synthetic life-forms available to those who can afford them, including snakes, fish, and the owl that Deckard, the main character, sees when he visits the Tyrell Corporation. The replicant, however, is the epitome of robotic life.

Genetic engineers design replicants and other robotic life forms from a combination of organic and synthetic materials. Though they are designed and built as machines, to some extent they grow organically.

Tyrell's Nexus series replicants, brought to market after the turn of the twenty-first century, blur the line between humans and robots. The most recent model, the Nexus 6, is stronger and smarter than the engineers who created it. Nexus 6 androids are almost indistinguishable from humans, except for one major difference—they were designed to have no emotions. Because of their sophistication, however, Tyrell's engineers expected replicants to develop emotions on their own. The Nexus 6 developers built in a four-year life span to counteract the androids' ability to adapt.

However, this plan was only marginally successful. After a bloody uprising off-world, replicants were made illegal on Earth—but that doesn't stop them from trying to reach the planet. Unfortunately, the six that jumped the shuttle for Earth are all trained for combat, which makes them particularly dangerous. One of them died in an electrical field as the group attempted to break into the Tyrell Corporation. We know the identities of only four more:

- Roy Batty, a combat model for the colonization defense program.
- Pris, a standard pleasure model for military use.
- Zhora, trained for political homicide.
- Leon, a combat model.

What they want is simple: a longer life—more time than the four years they've been given. As with most self-aware, living beings, they don't want to die.

Despite the prohibition on replicants, these six are not the only replicants on Earth. Rachel, who works at the Tyrell Corporation, is an experimental model even more sophisticated than the Nexus 6. It has a set of implanted memories, based on those of Tyrell's niece. Fake memories, Tyrell believes, make it easier for the androids, who are less experienced emotionally, to cope with their emotions. Having implanted memories will also make replicants easier for humans to control.

With the hazards replicants pose to humankind, why would the Tyrell Corporation continue to make them? One reason is that they still sell well among people off-world. The company also wants to improve upon its work, trying to achieve its motto: "More human than human."

Escaped replicants created the need for a whole new type of law enforcement. Next, we'll learn more about blade runners and what makes them tick.

BREAKING THE LAWS

Because they're willing to take human lives to realize their goals, replicants in *Blade Runner* break Isaac Asimov's three laws of robotics:

* A robot may not injure a human being or, through inaction, allow a human being to come to harm.
* A robot must obey the orders given it by human beings except where such orders would conflict with the First Law.
* A robot must protect its own existence as long as such protection does not conflict with the First or Second Law.

⚡ What's a Blade Runner?

A blade runner is a police officer charged with killing, or "retiring," replicants. Of course, blade runners don't want to take a human life mistakenly, so they test their targets first. The Voigt-Kampff test measures involuntary biological reactions. First, the blade runner sets up his equipment, which monitors physical signals like muscle movement in the subject's eye.

Once everything is ready and the subject sits still, the

blade runner asks questions designed to touch on the subject's morals—questions that should produce a noticeable emotional response. The machine works a little like a lie detector, measuring subtle changes in a subject's blush response and eye motion.

Since real animals are extremely scarce, blade runners ask questions that involve hurting or killing animals, which would produce a detectable emotional response in humans but not in replicants. At least, that's the theory.

HUMAN OR REPLICANT?

The Voigt-Kampff test can identify Nexus 6 replicants within twenty to thirty cross-referenced questions. But more sophisticated replicants may pose a greater challenge. It takes more than one hundred questions to identify Tyrell's assistant Rachael as a replicant. With its implanted memories, Rachael truly believes that it is human.

Once the blade runner identifies a replicant, the android is sure to try to defend its own life at all costs. Replicants can be aggressive killers, so a blade runner's job is extremely dangerous. So why do they take on the job? Mostly, it seems that no one else wants to—even Deckard quit at one point. His old supervisor, Harry Bryant, forces him back into service, though, when Bryant's best blade runner, Dave Holden, is shot by one of the escapees, and the other four killers are still on the loose. Deckard is simply Bryant's best chance of taking down the four "skin jobs" before they embarrass him and the police department.

Blade Runner is set in a futuristic world, but the idea of androids that are virtually indistinguishable from human beings isn't all that far-fetched. Next, we'll look at some of today's uncannily humanoid robots.

WOULD A CYLON PASS THE VOIGT-KAMPFF?

Chances are, one of the numbered Cylon models from Battlestar Galactica would pass the Voigt-Kampff test. Their highly advanced systems of these replicants allow them to hide undetected among human colonial citizens, even upon medical examination. If their advanced biology allows them to reproduce with humans, they can probably control their eye muscles during a Voigt-Kampff examination.

⚡Real Replicants: Today's Androids

Engineers have been working on androids for decades, but you can't visit your local department store to buy an android butler for your home. Yet.

For many years, Honda has been working on a sophisticated android, the Advanced Step in Innovative Mobility, or ASIMO. It can walk by itself and even climb stairs. As advanced as it is, though, ASIMO's artificial intelligence is somewhat limited. It can recognize faces and voices and is capable of doing helpful tasks around the home, such as turning on the lights and carrying items.

Repliee Q1Expo is an android developed by professor Hiroshi Ishiguro of Osaka University. Like the replicants in *Blade Runner*, Repliee Q1Expo closely resembles a human. It has soft silicone skin, eyelashes that move, and a set of forty-two actuators in its upper body that give it a range of motion. It even has motors that make it appear to "breathe."

Ishiguro didn't stop there. In 2007, he introduced the world to…himself. Actually, it was his doppelgänger, an android named Geminoid that looks just like him. Ishiguro can see through its eyes and talk through its internal speaker. Like Repliee Q1Expo, its chest rises and falls in breathing motions.

Don't be too surprised if you find Geminoid a little creepy. Roboticist Masahiro Mori theorizes that we're initially drawn to robots that seem more human-like, but once they get too human, we start to become repulsed. He wrote about this "uncanny valley" theory for the magazine *Energy* in 1970.

You might be surprised at how well a replicant can fool people into believing it's really a human. In fact, a question that has plagued *Blade Runner* fans for years is whether Deckard himself is a replicant or not. Either way, as we begin to develop more and more human-like robots, we will need to be very careful about whether we are crafting friends or foes—and we will have to make sure they never use whatever powers or abilities we give them against us, themselves, or anything else.

OTHER ANDROIDS: TV AND FILM

* Maria (*Metropolis*).
* The Cylons (*Battlestar Galactica*).
* C3PO (*Star Wars*).
* Data (*Star Trek*).
* Buffy Bot (*Buffy the Vampire Slayer*).
* Twiki (*Buck Rogers in the 25th Century*).
* Fembots (*Austin Powers*).
* VICI (*Small Wonder*).

TOP FIVE SCI-FI WEAPONS THAT MIGHT ACTUALLY HAPPEN

We've looked at some pretty cool upcoming super-powers in the world of warfare and robots, and as we saw in the first part of this book, the science-fiction and comic genres have amassed an impressive arsenal of weapon ideas over the years. From the phasers and red matter bombs of *Star Trek* to the lightsabers of *Star Wars*, our books, movies, and comics are loaded with a vast array of organic, nanotech, gravity, and energy weapons. But how much science is there to all of this? And just what kind of sci-fi heat will the soldiers of the future really be packing?

In this section, we'll look at and recap five far-out approaches to defending ourselves against destruction and mayhem—or wreaking havoc against future enemies. So check your crysknives, chainswords, and Klingon bat'leths at the door and take a quick glance into the future.

1. Powered Armor

Want to lift 300 pounds (136 kilograms), but you're not quite Schwarzenegger? DARPA and the Raytheon Company have made a robotic suit that makes people super strong. It multiplies lifting strength up to twenty times, and the U.S. Army hopes to use it for heavy lifting in the field. We've already looked at this a bit in the sections on body armor and liquid body armor, but there's another really cool aspect to body armor: powered armor that gives you superhuman strength.

This aluminum suit acts as an exoskeleton and fits almost like a jacket studded with position and motion sensors. Once the sensors feel the wearer's arm move, the suit follows through with its own hydraulics system. Both endurance and strength get a boost. The biggest drawback so far is the battery, which needs frequent recharging, and the 150-pound (68-kilogram) suit's weight, which makes it hard to drag around.

Over the years, powered armor has become a science-fiction mainstay, from the battle-hardened warriors of Robert Heinlein's 1959 novel *Starship Troopers* to the anime battlefields of *Mobile Suit Gundam* and the radiated wastes of *Fallout 3*. If an imagined future has soldiers in it, chances are they're outfitted in terrifying steel exoskeletons. No word on whether DARPA and Raytheon will seek inspiration from the space marines of *Warhammer 40000* for their next iteration of powered armor.

Are you not one to fight with brute force? With our next weapon (of sorts), you could change history instead.

⇄ 2. Time Travel

Time travel isn't a weapon, necessarily, but neither is a B-52 Stratofortress. The latter is a bomber, a delivery system for weapons. Likewise, if our future operates the way it did in the *Terminator* films, we might use time travel to assassinate future military leaders before they rise to power. Or perhaps we'll actually wage full-blown wars across four dimensions, such as the Time War waged between the Time Lords and the Daleks of TV's *Doctor Who*.

Real time travel is much trickier than movies would lead us to believe, but it is within the laws of physics. Einstein taught us that time slows down—or to think of it visually, stretches out—when you travel close to the speed of light. You only need an airplane to notice the effect. In a famous experiment, physicists synchronized five atomic clocks, then kept one on the ground and put the rest on two very fast airplanes (one heading eastward, the other westward). After the airplanes landed, their clocks measured less time than the ground clock. The difference was tens to hundreds of nanoseconds. If the clocks were people, the airplane clocks would be younger than the ground clock.

Of course, nanoseconds don't interest us. We want to meet ourselves as children or old folks. Serious time travel requires more than an airplane. It requires us to play around with black holes, wormholes, or cosmic strings, all of which we're still studying as phenomena.

We can't really say whether we'll ever be able to instantly go forward or backward in years (except by aging), but the physical framework is real enough for it to make our list.

Next, we'll play with more physics.

⇅ 3. Teleportation

Sure, it didn't vaporize people like thalaron radiation did. But when Captain Kirk and the *Star Trek* crew needed to escape, they jumped into the teleporter. As we learned earlier, Kirk just stood on a platform, and this wickedly cool machine mapped every atom in his body. It sent information about the atoms by light waves (just like the Internet sends information by radio waves) to a new place. In the new place, machinery received the information about Kirk's atoms and rebuilt Kirk.

As we saw, in real life, scientists have teleported objects: a photon and a laser beam. But there are big barriers to teleporting humans. First, as Kevin Bonsor points out in "How Teleportation Works," we'd need to find and describe all 10^{28} atoms in the body, which we can't do. Second, to reassemble the person, we'd need to put each atom in the right place and make sure it had the right properties. A tiny imprecision could be deadly.

This means no beaming your soldiers behind enemy lines, much less "telefragging" someone by teleporting things directly into his physical coordinates.

If you'd like to keep your atoms, read on for a weapon on a much larger scale.

⚡4. Asteroid Defense (or Offense)

Asteroids may call to mind lots of images: Earth shrouded in dust, dinosaurs dying, crowds running, Aerosmith singing "I Don't Want to Miss a Thing." Asteroids have made plenty of disaster-movie plots, all of which go like this: don't panic, but a giant rock is headed for Earth.

Here's how it could go: either a comet (if an outer planet's gravity pulled one closer to Earth) or an asteroid could cross Earth's orbit or pass nearby. NASA watches for these near-Earth objects and plans to find most of the ones 1 kilometer (0.6 miles) in diameter and larger, as well as learn which ones might collide with Earth. Why 1 kilometer? Anything with that diameter or bigger could do terrible damage.

THE HOLLYWOOD EFFECT

Movies like *Deep Impact*, *Armageddon*, and *Asteroid* get a lot wrong. For example, in *Armageddon* a comet collides with an asteroid, knocking an asteroid the size of Texas toward Earth. No asteroid in our solar system is that big, and no comet could knock an asteroid that big at us.

Real plans to divert an asteroid are more like sketches than ready-to-use plans. With ten years of warning and a medium-sized asteroid, we might plant a nuclear bomb on or near the asteroid. With twenty years of warning and a small asteroid, we might collide an unmanned spacecraft with the asteroid to verify its location and slow it and divert it.

But as Carl Sagan points out in *Pale Blue Dot*, if an asteroid were headed toward Earth anyway, and we had mastered how to intercept and deflect it, nations could try to knock the asteroid at one another. So keeping the peace on our own planet is always a priority, too.

The next weapon is as gentle as a butterfly.

⇏ 5. Insect Cyborgs

It's an old idea to train animals for use in war. We've trained bees as bomb-sniffers and used dolphins to patrol our ports. The weaponized insects of *Aeon Flux* may be a ways off, but making animals into machines is already a reality. Working under DARPA, researchers have made real moths and beetles into remote-controlled robo-bugs. In the future, the insects may carry cameras or chemical sensors into the field.

The engineering starts when the bugs are larvae. At this time, you can implant wires into the bugs, and their bodies grow healthily around them. Their nerves, muscles, and brains intertwine electrically with the implants.

That intermingling helps in bug control because moths and beetles operate on reflexes. Because entomologists know which patterns to send to which nerves to trigger a behavior, they can use electrodes to hijack the bugs. For example, stimulating the neck muscles makes the bug circle left or right. By implanting a tiny radio receiver on the moth's back, researchers can control it wirelessly from a joystick.

In mechanizing moths, researchers encountered the hilarities you'd expect. Tobacco moths, the kind used in the experiments, must shiver for five minutes to warm their flight muscles before they fly. Not wanting to wait, researchers implanted heaters to warm the muscles. And when carrying so much metal, the moths can't fly their normal range of kilometers without getting exhausted. Not to worry. The researchers hung the moths from helium balloons. For obvious reasons, balloons wouldn't work for spying missions.

NOW YOU SEE ME, NOW YOU DON'T: HOW INVISIBILITY CLOAKS WORK

Weapons and armor are fine, but they really only protect you from physical attacks or in the heat of battle. And how many of us ordinary people actually find ourselves in those situations on a regular basis? What we need is something that can help us slip away from everyday embarrassing situations or let us duck that annoying coworker or crazy boss in the hall. We need Harry Potter's invisibility cloak!

Admit it. You'd love to own an invisibility cloak. Utter an embarrassing faux pas at a party? Just throw on your magical garment and vanish from the snooty gaze of your fellow partygoers. Want to hear what your boss is really saying about you? Stroll right into his or her office and get the goods.

Such fantastic fashion accessories have become ridiculously standard in the world of science fiction and fantasy. Everyone, from boy wizards to intergalactic safari hunters, has at least one invisible blouse in their wardrobe, but what about us poor saps in the real world?

Well, Muggles, science has some good news for you:

invisibility cloaks are a reality. The technology is far from perfect, but if you'll step into our high-tech boutique of vanishing apparel, we'll guide you through your invisibility cloak options.

* First up, we'll look at some wonderful carbon nanotube fashions—fresh from the University of Texas at Dallas's NanoTech Institute collection. We've already seen this technology being tested for various body armor uses, and it's inspired by the same natural phenomena responsible for desert mirages. Heated via electrical stimulation, the sharp temperature differential between the cloak and the surrounding area causes a steep temperature gradient that bends light away from the wearer. The catch: wearers must love water and be able to fit inside a petri dish.

* Or perhaps you'd prefer something made from metamaterials. These tiny structures are smaller than the wavelength of light. If properly constructed, they guide rays of light around an object—much like a rock diverting water in a stream. For now, however, the technology only works in two dimensions and only comes in the ultra-petite size of 10 micrometers across.

* If you're more into retro fashion, there's also the optical camouflage technology developed by scientists at

the University of Tokyo. This approach works on the same principles as the blue screen used by TV weather forecasters and Hollywood filmmakers. If you want people to see through you, then why not just film what's behind you and project it onto your body? If you travel with an entourage of videographers, this may be the cloak for you.

Ready to try some of these fashions on for size?

The Mirage Effect: Carbon Nanotubes

First, let's try this carbon nanotube invisibility cloak on for size and experience the wonders of the mirage effect.

You're probably most familiar with mirages from tales of desert wanderers who glimpsed a distant oasis, only to discover it was just a mirage—no miraculous lake of drinking water, merely more hot sand.

The hot sand is key to the mirage effect (or photothermal deflection), as the stiff temperature difference between sand and air bends, or refracts, light rays. The refraction swings the light rays up toward the viewer's eyes instead of bouncing them off the surface. In the classic example of the desert mirage, this effect causes a "puddle" of sky to appear on the ground, which the logical (and thirsty) brain interprets as a pool of water. You've probably seen similar effects on hot roadway surfaces, with distant stretches of the road appearing to gleam with pooled water.

In 2011, researchers at the University of Texas at Dallas NanoTech Institute managed to capitalize on this effect. They used sheets of carbon nanotubes, sheets of carbon wrapped up into cylindrical tubes. Each page is barely as thick as a single molecule, yet it is as strong as steel because the carbon atoms in each tube are bonded incredibly tightly. These sheets are also excellent conductors of heat, making them ideal mirage-makers.

In the experiment, the researchers heated the sheets electrically, which transferred the heat to the surrounding area (a petri dish of water). This caused light to bend away from the carbon nanotube sheet, effectively cloaking anything behind it with invisibility.

Needless to say, there aren't many places you'd want to wear a tiny, super-heated outfit that has to stay immersed in water, but the experiment demonstrates the potential for such materials. In time, the research may enable not only invisibility cloaks but also other light-bending devices—all of them with a handy on-off switch.

Metamaterials: Bending Light Waves

Next, let's slip into an invisibility cloak made from metamaterials.

Metamaterials offer a more compelling vision of invisibility technology, without the need for multiple projectors and cameras. First conceptualized by Russian physicist Victor Veselago in 1967, these tiny, artificial structures are smaller than the wavelength of light (they have to be to divert them) and exhibit negative electromagnetic properties that affect how an object interacts with electromagnetic fields.

Natural materials all have a positive refractive index, and this dictates how light waves interact with them. Refractivity stems in part from chemical composition, but internal structure plays an even more important role. If we alter the structure of a material on a small enough scale, we can change the way it refracts incoming waves—even forcing a switch from positive to negative refraction.

Remember, images reach us via light waves. Sounds reach us via sound waves. If you can channel these waves around an object, you can effectively hide it from view or sound. Imagine a small stream. If you stick a tea bag full of red dye into the flowing water, its presence would be apparent downstream, thanks to the way it altered the water's hue, taste, and smell. But what if you could divert the water around the tea bag?

In 2006, Duke University's David Smith took an earlier theory posed by English theoretical physicist John Pendry and used it to create a metamaterial capable of distorting the flow of microwaves. Smith's metamaterial fabric consisted of concentric rings containing electronic microwave distorters. When activated, they steer frequency-specific microwaves around the central portion of the material.

Obviously humans don't see in the microwave spectrum, but the technology demonstrated that energy waves could be routed around an object. Imagine a cloak that can divert a third grader's straw-fired spitball, move it around the wearer, and allow it to continue on the other side as if its trajectory had taken it, unopposed, straight through the person in the cloak. Now how much more of a stretch would it be to divert a rock? A bullet?

Smith's metamaterials proved the method. The recipe to invisibility lay in adapting it to different waves.

NANOTECHNOLOGY TO THE RESCUE

Metamaterials, a creation of science, don't occur naturally. To create the minute structures required to redirect electromagnetic waves, scientists employ nanotechnology.

Metamaterials: Invisible Tanks

In 2007, the University of Maryland's Igor Smolyaninov led

his team even farther down the road to invisibility. Incorporating earlier theories proposed by Purdue University's Vladimir Shaleav, Smolyaninov constructed a metamaterial capable of bending visible light around an object.

A mere 10 micrometers wide, the Purdue cloak uses concentric gold rings injected with polarized cyan light. These rings steer incoming light waves away from the hidden object, effectively making it invisible. Chinese physicists at Wuhan University have taken this concept into the audible range, proposing the creation of an acoustic invisibility cloak capable of diverting sound waves around an object.

For the time being, metamaterial invisibility cloaks are somewhat limited. They're not only small; they're limited to two dimensions—hardly what you'd need to vanish into the scenery of a Victor Veselago war zone. Plus, the resulting cloak would weigh more than even a full-grown wizard could hope to lug around. As a result, the technology might be better suited to applications such as hiding stationary buildings or vehicles, such as a tank.

Optical Camouflage: Altered Reality

Ready to slip into some old-school optical camouflage fashions?

This technology takes advantage of something called augmented-reality technology—a type of technology first pioneered in the 1960s by Ivan Sutherland and his students at Harvard University and the University of Utah.

Optical camouflage delivers a similar experience to Harry Potter's invisibility cloak, but using it requires a slightly complicated arrangement. First, the person who wants to be invisible (let's call him Harry) dons a garment that resembles a hooded raincoat. The garment is made of a special material that we'll examine more closely in a moment.

Next, an observer (let's call him Professor Snape) stands before Harry at a specific location. At that location, instead of seeing Harry wearing a hooded raincoat, Snape sees right through the cloak, making Harry appear to be invisible. And if Snape stepped to the side and viewed Harry from a slightly different location? Why, he'd simply see the boy wizard wearing a silver garment. Scowls and detentions would likely follow. Luckily for Harry, his fictional cloak affords 360-degree protection.

Optical camouflage doesn't work by way of magic. It works by taking advantage of something called augmented-reality technology—a type of technology first pioneered by Ivan Sutherland, as mentioned earlier.

Augmented-reality systems add computer-generated information to a user's sensory perceptions. Imagine, for example, that you're walking down a city street. As you gaze at sites along the way, additional information appears to enhance and enrich your normal view. Perhaps it's the day's specials at a restaurant or the show times at a theater or the bus schedule at the station. What's critical to understand is that augmented reality is not the same as virtual reality. While virtual reality aims to replace the world, augmented reality merely tries to supplement it with additional, helpful content. Think of it as a heads-up display for everyday life.

Most augmented-reality systems require a user to look through a special viewing apparatus to see a real-world scene enhanced with synthesized graphics. They also call for a powerful computer. Optical camouflage requires these things as well, but it also necessitates several other components. Here's everything needed to make a person appear invisible:

* A garment made from highly reflective material.
* A digital video camera.
* A computer.
* A projector.
* A special, half-silvered mirror called a combiner.

In the next section, we'll look at each of these components in greater detail.

⚡Optical Camouflage: Invisibility Cloak Components

All right, so you have your video camera, computer, projector, combiner, and wondrous reflective raincoat. Just how does augmented-reality technology turn this odd shopping list into a recipe for invisibility?

First, let's take a closer look at the raincoat: it's made from retro-reflective material. This high-tech fabric is covered with thousands and thousands of small beads. When light strikes one of these beads, the light rays bounce back exactly in the same direction from which they came.

To understand why this is unique, look at how light reflects off other types of surfaces. A rough surface creates a diffused reflection because the incident (incoming) light rays scatter in many different directions. A perfectly smooth surface, like that of a mirror, creates what is known as a specular reflection—a reflection in which incident light rays and reflected light rays form the exact same angle with the mirror surface.

In retro-reflection, the glass beads act like prisms, bending the light rays by refraction. This causes the reflected light rays to travel back along the same path as the incident light rays. The result: an observer situated at the light source receives more of the reflected light and therefore sees a brighter reflection.

EVERYDAY RETRO-REFLECTION

Retro-reflective materials are actually quite common. Traffic signs, road markers, and bicycle reflectors all take advantage of retro-reflection to be more visible to people driving at night. The movie screens found in most modern commercial theaters also take advantage of this material because it allows for high brilliance under dark conditions. In optical camouflage, the use of retro-reflective material is critical because it can be seen from far away and outside in bright sunlight—two requirements for the illusion of invisibility.

Optical Camouflage: More Invisibility Cloak Components

For the rest of the setup, the video camera needs to be positioned behind the subject to capture the background. The computer takes the captured image from the video camera, calculates the appropriate perspective, and transforms the captured image into the picture that will be projected onto the retro-reflective material.

The projector then shines the modified image on the garment by shining a light beam through an opening controlled by a device called an iris diaphragm. This diaphragm is made of thin, opaque plates. Turning a ring changes the diameter of the central opening.

TINY OPENING, LARGE DEPTH OF FIELD

For optical camouflage to work properly, the opening must be the size of a pinhole. Why? This ensures a larger depth of field so that the screen (in this case, the cloak) can be located any distance from the projector.

Finally, the overall system requires a special mirror to both reflect the projected image toward the cloak and let light rays bouncing off the cloak return to the user's eye. This special mirror is called a beam splitter, or a combiner—a half-silvered mirror that both reflects light (the silvered half) and transmits light (the transparent half).

If properly positioned in front of the user's eye, the combiner allows the user to perceive both the image enhanced by the computer and light from the surrounding world. This is critical because the computer-generated image and the real-world scene must be integrated fully for the illusion of invisibility to seem realistic. The user has to look through a peephole in this mirror to see the augmented reality.

Now, let's take a look at how this whole system comes together.

⇌ Optical Camouflage: The Complete Invisibility System

Let's put all of these components together to see how the invisibility cloak appears to make a person transparent. Once a person puts on the cloak made with the retro-reflective material, here's the sequence of events:

1. A digital video camera captures the scene behind the person wearing the cloak.

2. The computer processes the captured image and makes the calculations necessary to adjust the still image or video so it will look realistic when it is projected.

3. The projector receives the enhanced image from the computer and shines the image through a pinhole-sized opening onto the combiner.

4. The silvered half of the mirror, which is completely reflective, bounces the projected image toward the person wearing the cloak.

5. The cloak acts like a movie screen, reflecting light directly back to the source, which in this case is the mirror.

6. Light rays bouncing off the cloak pass through the transparent part of the mirror and fall on the user's

eyes. Remember that the light rays bouncing off the cloak contain the image of the scene that exists behind the person wearing the cloak.

The person wearing the cloak appears invisible because the background scene is being displayed onto the retro-reflective material. At the same time, light rays from the rest of the world are allowed to reach the user's eye, making it seem as if an invisible person exists in an otherwise normal-looking world.

⚡ Optical Camouflage: Real-World Invisibility Applications

The words "invisibility cloak" tend to summon images of fantastic adventure, magical espionage, and otherworldly deception. The actual applications for optical camouflage, however, are far less out there. You can forget hiding your Romulan starship or hanging out in the lady wizards' dormitory, but that doesn't mean there aren't a number of viable uses for the technology.

* For instance, pilots landing a plane could use this technology to make cockpit floors transparent. This would enable them to see the runway and the landing gear simply by glancing down at the floor (which would

display the view from the outside of the fuselage).

* Similarly, drivers wouldn't have to deal with mirrors and blind spots. Instead, they could just "look through" the entire rear of the vehicle.

* The technology even boasts potential applications in the medical field, as surgeons could use optical camouflage to see through their hands and instruments for an unobstructed view of the underlying tissue.

Interestingly enough, one possible application of this technology actually revolves around making objects more visible. The concept is called mutual telexistence and essentially involves projecting a remote user's appearance onto a robot coated in retro-reflective material. Say a surgeon was operating on a patient via remote-control robotic surgery. Mutual telexistence would provide the human doctors assisting the procedure with the perception that they're working with another human instead of a machine.

Right now, mutual telexistence is science fiction, but scientists continue to push the boundaries of the technology. For example, pervasive gaming is already becoming a reality. Pervasive gaming extends gaming experiences out into the real world, whether on city streets or in remote wilderness. Players with mobile displays move through the world while sensors capture information about their environment, including their location. This information delivers a gaming experience that changes according to where users are and what they are doing.

So we may still be far away from having invisibility cloaks we can just switch on or off any time we like, but they could be a possibility in the future! And what a great superpower that would be: being able to be visible or invisible whenever we want. To wrap up our jaunt through the crazy cool science of being super, let's take a look at some superpower-granting gadgets that we wish really existed.

WISHFUL THINKING: TEN SCI-FI GADGETS WE WISH ACTUALLY EXISTED

Science fiction can be a powerful genre. Apart from often imparting deep social commentary, science fiction has given us other gifts: amazing inventions that we'd love to possess. Some sci-fi gadgets became reality. As we learned earlier, *Star Trek* introduced the concept of a universal translator—a gadget capable of making communication possible across language barriers. Today, you can use a smartphone and Google Translate to have a conversation with someone else even if you don't share a common language. There are thousands of examples of real-world gadgets and inventions that were once just the stuff of dreams.

But not all gizmos and doodads from sci-fi are available at the local retail store. To close, we're going to recap ten gadgets introduced in sci-fi that we're just dying to get our hands on. Some you've seen earlier in this book; some you haven't. All should be real. If only we ordinary guys could conjure such great things into existence, as so many of our favorite superheroes and sci-fi legends have.

⚡1. The Hoverboard

Back to the Future 2 had a lot to live up to. The first film was a runaway success. It introduced the flux capacitor—a component we'll visit a little bit later. It also created a market for DeLorean cars, a vehicle that by 1985 was on the road to oblivion. And while you might argue that the second film lacks the charm and pacing that was present in the original movie, it did capture our imaginations with the hoverboard.

Simply put, a hoverboard is a skateboard without the wheels. It defies gravity, allowing the rider to zoom above the ground. To turn on a hoverboard, you simply lean as if you were on a normal skateboard. As we learn in the film, hoverboards don't work on water unless you've got power.

How do they work? It beats us! The film never really attempts to explain what makes hoverboards tick. We know we'd love to own one and swoop around the office. Shortly after the film hit theaters, a myth circulated that the hoverboards in the film were real products—they even had the Mattel logo on them. But the myth said that consumer groups and concerned parents pressured Mattel to pull hoverboards from production out of fear that the boards would cause countless injuries.

In truth, there never were any working hoverboards—all those effects came from movie magic. But we're still holding out hope that one day we'll get to glide along with Huey Lewis music blasting in the background.

⩲ 2. The Neuralizer

For those of us who suffer from foot-in-the-mouth disease, no gadget would be handier than the neuralizer. An important tool in the arsenal for the Men in Black, this gadget lets you zap away the memories of those who stare at the flashing red light. With the click of a button and a few soothing words, you can wipe out a memory and replace it with something else.

For the characters in *Men in Black*, this device allowed human agents to meet, negotiate, or combat aliens without alerting the entire Earth that we are not alone. But in the HowStuffWorks.com office, we'd probably use this in other ways. Need a little more time on that deadline? Just zap the site director and say that the assignment is due next week. Accidentally spill coffee on the general manager? A quick zap and the suggestion that one of the Stuff You Should Know guys did it, and you're good to go.

Zapping people to alter memories might not be the most responsible option. Maybe it's a good thing the neuralizer doesn't really exist. But if you see a HowStuffWorks .com writer walking around wearing sunglasses, you might want to avert your eyes—just in case.

⚡3. The Lightsaber

Putting aside the legitimate argument that the *Star Wars* series is really more of a fantasy than science fiction, we come once again to the lightsaber. It's an elegant weapon from a more civilized age. The final part of a Jedi knight's training is the construction of his or her personal lightsaber. The films taught us that these magical swords could cut through nearly anything and were capable of deflecting blaster fire. Plus they make that really cool voom-whoosh sound.

LIGHTSABER LITERATURE

If you explore the expanded universe—that includes the various novels, video games, comic books, and other media that relate to *Star Wars* but aren't part of the official story—you'll learn that a lightsaber consists of a handle, a power source, and some crystals. The crystals give the lightsaber its color as well as other attributes. Those who use the light side of the Force tend to rely on crystals they find in natural settings like caves and caverns. Dark-side Force users tend to use synthetic crystals, which always seem to give a lightsaber an ominous red glow.

While we don't foresee the need to put a lightsaber to any sort of combat use here at HowStuffWorks.com, we admit it would be really handy for yard work. With a couple of quick swipes, you could cut down trees, bushes, and any plastic pink flamingoes that are between you and the perfectly manicured lawn.

4. The Electronic Thumb

Legend has it that Douglas Adams thought up the idea for *The Hitchhiker's Guide to the Galaxy* while lying in a field recovering from drinking a bit too much during a trip through Europe. It wasn't uncommon for students and other travelers to hitch a lift now and then as they crisscrossed the continent, visiting new cities and phoning home for more money. What if, thought Adams, the same thing happened on a universal scale? He constructed a tale of a befuddled human named Arthur Dent and an alien in disguise with the vehicular moniker of Ford Prefect, and the rest is history.

But how do you hitch a ride with an alien? You use an Electronic Thumb. Adams explains that there is a communications channel called the sub-ether network. The electronic thumb taps into this network and signals nearby spaceships to hitch a lift. Adams wrote multiple versions of his story, and no two are exactly alike. It's not entirely clear whether the thumb requires the spaceship's driver to give permission before the hitchhikers zap aboard using a matter transference beam.

It's true that for the electronic thumb to really be useful we'd need to have some aliens flying around first. But even if there aren't any bug-eyed monsters in the nearby galaxies, it would still make a lovely paperweight.

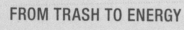 5. Mr. Fusion

FROM TRASH TO ENERGY

Why are we going back to *Back to the Future* for this one, you might be thinking, especially since it wasn't covered in this book. The trilogy introduced lots of cool gadgets: time machines, flying skateboards, and self-tying shoes are just a few examples. But Mr. Fusion could revolutionize everything.

It's a throwaway visual gag at the end of the first *Back to the Future* film—Doc hurriedly sorts through Marty's garbage can, pulling out banana peels and beer. He feeds it into the Mr. Fusion port on the back of the time machine. The big joke is that this relatively tiny device can generate the awesome power—1.21 gigawatts' worth—that the flux capacitor needs in order to make time travel possible. Throughout the entire film we've watched Marty and Doc try to harness lightning to get Marty back to 1985, and by 2015 the same power can be generated by an off-the-shelf appliance.

But imagine how different our world would be if Mr. Fusion were real. We could generate all our power needs just by feeding in some garbage. A couple of nuclear reactions later, we'd have plenty of juice to run our homes and vehicles. It solves recycling problems and energy conservation all in one fell swoop! Sure, there are lingering concerns about using a nuclear reactor in such a casual way, but without risk there's no reward, right?

6. Iron Man's Armor

Many comic-book superheroes possess amazing powers. A few, like Batman or Iron Man, are relatively normal human beings who rely on their training and gadgets to get the upper hand on villains. Iron Man's suit is the Swiss Army knife of the super-gadget world. It can fly; it's impervious to most forms of damage; and it features repulsor beams that can blast holes in masonry.

IRON MAN'S CHANGING WARDROBE

Throughout the history of the Iron Man's appearances in comics, television shows, and films there have been many versions of the armor. Some are large and bulky, resembling a tank more than anything else. Others are sleek—one suit even had the ability to fold up into a suitcase. How Stark managed to carry around a suit like that without his back giving out remains something of a mystery.

With an Iron Man suit, casual Friday at the HowStuffWorks.com office would take on a new meaning. And it sure would be handy to step out onto our balcony and fly off to grab lunch—there'd be no need to wait on those pesky elevators!

7. Time Machines

While the neuralizer from *Men in Black* would be handy for making folks forget the last few moments, what happens when you make a major mistake? That's when you hop into your handy-dandy time machine and risk introducing a paradox that could rip apart the very fabric of time and space in order to prevent yourself from an embarrassing situation.

Sure, you might encounter a younger version of yourself or somehow set into motion a series of events that will prevent your own birth, but that's half the fun!

As we've seen in this book, time machines come in all shapes, sizes, and styles. You could have a living creature that inhabits a multidimensional construct like the TARDIS in *Doctor Who*. You might plop down on a comfy chair and manipulate dozens of levers and dials like the time machine in H.G. Wells's famous story. Or maybe you want to give that flux capacitor a real workout and zoom into time at 88 miles per hour (141.6 kilometers per hour) with the famous DeLorean from the *Back to the Future* films. No matter what your sense of personal style, there's a time machine out there for you.

OBEY TIME: IT'S THE LAW

Is time travel possible? It's still a matter of debate. If the universe behaves the way Einstein predicted, time travel may be impossible. But scientists in 2011 were surprised when they discovered that neutrinos—sub-atomic particles—appeared to travel faster than the speed of light. While that may have been an error in measurement, if the neutrinos really did exceed the universal speed limit, time travel might just be a reality on a sub-atomic scale.

⇌ 8. The Transporter

We looked at this in detail earlier, but it's worth revisiting because a transporter would be so darn useful! Do you live in one of the worst cities for commuters? If you do, you know the frustration of having to sit in traffic that's at a standstill for what feels like hours on end. In the worst of these cities, you might spend more than seventy hours every year in traffic delays. But what if you could skip the commute entirely? That's why a *Star Trek* transporter needs to be invented ASAP. Well, for that reason and a lot of others.

TRANSPORTER 101

To recap quickly, the transporter was the device that could dematerialize you, shoot you across vast distances, and reassemble you at your destination. You didn't even need two of them—a single transporter could plop you down from a starship to a planet below and scoop you back up again once your mission was over—or when enough guys in red shirts had shuffled off the mortal coil. As we learned earlier, a popular term for the act of being transported is "beaming."

In *Star Trek* lore, the transporter originally only moved non-biological cargo. Disassembling atoms of a non-living target and putting them back together on a ship isn't as scary as the thought of having all your own atoms ripped apart and squished back together. Some famous doctors in the *Star Trek* universe took great pains to avoid having to use the transporters, preferring shuttles to dematerialization. Still, it beats an hour-long commute through gridlocked traffic!

9. The Replicator

Star Trek also contributed another gadget worth including in this list: the replicator. As you might guess from its name, the replicator can create stuff as long as it knows what that stuff is made of on a molecular level. If you have the molecular recipe for lasagna, the replicator can whip up a nice batch for you on the spot.

Could we ever build an actual working replicator? Scholars like K. Eric Drexler think it might be a possibility with nanotechnology. We may one day be able to construct machines that measure only a few billionths of a meter across—so small you can't even see them with a light microscope. These molecular machines could, in theory, assemble material one molecule at a time. With billions of these assemblers, you could create practically anything as long as you had the raw materials at hand. But before you throw away your microwave and ditch your oven, you should know that there are other scientists like Richard E. Smalley who think that fundamental barriers make devices like molecular assemblers a virtual impossibility.

For the time being, we have to make our hot Earl Grey tea the old-fashioned way. But we're still hoping nanotechnology brings the replicator to reality!

10. The Sonic Screwdriver

First, how cool of a name is that? There may be no gadget as versatile and useful as the Doctor's sonic screwdriver from the *Doctor Who* series. It can open (or engage) locks ranging from rusty old padlocks to digital keypads. It can reprogram computers and repair old wiring. In a pinch, you can use it as a weapon and knock people unconscious with it, or pair it with a power source to zap Daleks or Cybermen. What can't it do?

Well, it can't open anything that has a deadlock seal on it. What's that? It's a plot device designed to make it harder for the Doctor to escape a situation. In other words, a sonic screwdriver works in any situation except when it's not convenient to the plot. We'd love to have this kind of device. Most of the time, it will work perfectly. When it doesn't work, you know you're in a really important situation.

While the Doctor has had decades to become adept at

wielding the sonic screwdriver, his human traveling companions have also put it to use in a pinch. That gives us hope that this incredibly flexible tool would still work in the hands of a novice. We eagerly await the full-scale production of the device—no toolkit should be without one.

CONCLUSION

S o how does it feel to be super knowledgeable about being superhuman? If only we could actually *be* superhuman. Luckily, as we've learned, that possibility isn't totally far-fetched, and in some cases, superhuman abilities are not all that far off. We're already starting to create cloaking devices, universal translators, phasers, and tractor beams. Next up, time travel and teleportation! Of course, every superpower has a dark side, so we'll need to watch out for that. But we can look forward to a pretty awesomely super future.

QUIZ: THE ULTIMATE SUPERHERO QUIZ

So now that we've taught you about everything from lightsabers and replicants to warp speed and sonic screwdrivers, it's time to see what you've learned with the Ultimate Superhero Quiz. Our favorite sci-fi and comic-book heroes and villains are tested all the time, so now we're going to test you mortals. If you pass, you'll have super knowledge of the science of being super, and who knows, maybe you will be on your way to becoming a superbeing yourself! Check your answers in the back of the book on page 284.

1. **After extensive consideration, who did HowStuffWorks conclude would win in a fight between Superman and Gandalf the magician?**

    a. Gandalf

    b. Superman

    c. It would be a draw.

2. **What is the name of the X-Men superhero who can absorb ordinary people's strength and memories as well as mutants' superpowers?**

   a. Storm

   b. Wolverine

   c. Rogue

3. **Which one of the following criteria for antisocial personality disorder (APD) does Batman exhibit?**

   a. Reckless disregard for the safety of self or others

   b. Irritability or aggressiveness

   c. Insensitivity to pain

4. **What are Wolverine's claws made of?**

   a. Cadmium

   b. Adamantium

   c. Titanium

5. **What does the special memory fiber in Batman's cape allow him to do?**

   a. Communicate with the Batmobile

   b. Slowly fall

   c. Disappear

6. Which of the following *Star Trek* technologies actually has become a reality?

   a. Human teleportation

   b. The replicator

   c. Hypospray

7. True or False: Salamanders are the most complex species of animals capable of regenerating body parts.

   a. True

   b. False

8. True or False: The new superpower reality of liquid body armor is made from a Kevlar that has been soaked in a shear-thinning fluid to make give it a flexible, liquid layer and increased defensive strength.

   a. True

   b. False

9. What is the main test to determine if someone is a replicant in *Blade Runner*?

   a. A blood test

   b. A brain scan

   c. A muscle movement test

10. Which of the following technologies can be used to create an invisibility cloak?

a. Carbon nanotubes

b. Dematerialization

c. Magnetorheological fluids

ANSWERS TO THE QUIZZES

⚡The Ultimate Batmobile Quiz

1. **How much does the Batmobile weigh?**

    Answer: The Batmobile used in the film *Batman Begins* was a real vehicle. It weighed 5,000 pounds and even served as the pace car in a NASCAR race.

2. **Which of the following is not a Batmobile driving position?**

    Answer: Batman's ride has two driving settings: one for regular driving and one for jumping/flying. The Batmobile does not have an amphibious/swimming position. It is also powered by a jet engine and has stealth capabilities, thanks to a silent, electronic motor.

3. **What is model bashing?**

    Answer: Model bashing involves using parts from toy

cars, plastic models, metal and plastic tubes, and other parts to build a small version of the car you want to create in reality.

4. **The Batmobile's creator built the model for the real Batmobile on a _____ scale.**

    Answer: The Batmobile's creator built six models, all on a 1:12 scale, before he got the look and the shapes that he wanted for the actual design.

5. **After a scale model was designed, a full-size model of the Batmobile was built from:**

    Answer: The Batmobile was sculpted at full size out of Styrofoam. This helps determine actual proportions as well as the design for the body-panel molds and frame of the Batmobile.

6. **To play its role in the film *Batman Begins*, the actual Batmobile had to accelerate from 0 to 60 in:**

    Answer: For its role in the film, the Batmobile had to be able to accelerate from 0 to 60 in five seconds. It also had to reach speeds of over 100 mph, turn corners at high speeds, and land a successful jump of 30 feet without damage.

7. **In what city were the car scenes involving the Batmobile filmed for *Batman Begins*?**

Answer: All of the scenes involving the Batmobile in *Batman Begins* were filmed on the streets of Chicago, Illinois.

8. **The Batmobile's rear tires measured:**

Answer: The rear tires of the Batmobile measured 37 inches in diameter. They were off-the-shelf, 4x4 mud tires called Super Swampers, made by Interco.

9. **How much did each street-ready Batmobile used in filming cost to build?**

Answer: Each street-ready racing model of the Batmobile used in filming cost a quarter of a million dollars to build. Because the car had no side windows and many exposed parts, drivers trained for six months before filming scenes. Different types of Batmobiles were also used for different scenes, including a "flap version" and a "jet version."

10. Which of the following models was used in the scenes in which the Batmobile travels between buildings?

Answer: The miniature Batmobile wasn't actually that small; it was a 6-foot-long, 1:5 scale model of the Batmobile, complete with an electric motor drive. When the Batmobile flew through the air and across ravines or between buildings in the film, this scale model did the flying.

⚡ May the Force Be with You: A Striking *Star Wars* Quiz

1. True or False: The Sith always act in groups to increase the force of their attack.

   Answer: The Sith only operate in pairs: one master and one apprentice.

2. What is the name of the remote and uninhabited planet that was destroyed by the first test of the Death Star's laser cannon?

   Answer: Despayre was the planet demolished by the Death Star's cannon during its first test.

3. In the *Star Wars* series, what serves as the energy supply for a lightsaber?

   Answer: The diatium core creates the energy that powers a lightsaber.

4. Who did Anakin Skywalker have to defeat to become the most legendary of all Sith Lords?

   Answer: Darth Vader (then known as Anakin Skywalker) killed Darth Tyranus, went over to the dark side, and became Darth Vader, the most legendary of all Sith Lords.

5. True or False: According to *Star Wars*, not just anyone can tap into the Force and use it at their will.

   Answer: The Force energy flows through all living things, but certain individuals and creatures have a natural adeptness or Force sensitivity. When Force-sensitive individuals tap into this, they can unlock unimaginable power within.

6. The Rebel Alliance destroyed the Death Star for the first time during what battle?

Answer: The rebels first destroyed the Death Star during the Battle of Yavin in Episode 4. The battle was so named because of the hidden stronghold the rebels kept on the fourth moon of planet Yavin. The Imperial Force referred to the attack as the Yavin Massacre.

7.   Which of the following is not one of the six sectors within each of the zones on the Death Star?

Answer: The six sectors on the Death Star are: General, Command, Military, Security, Service, and Technical.

8.   True or False: The superlaser's power needs to be recharged between blasts.

Answer: The superlaser's power needs to be recharged between blasts, limiting it to only one planet-destroying beam per day.

9.   The Great Schism produced what distinct group?

Answer: Jedi fought Jedi in the one hundred-year war between the light and dark sides of the Force, known as the Great Schism. The dark Jedi were exiled to Korriban, where they enslaved and bred

with the native Sith population to eventually produce the Sith Lords.

10. **How did the Rebels finally defeat the Sith in the Galactic Civil War?**

Answer: The Rebels took down the shield generator
and destroyed the Death Star. Both the Galactic Civil
War and the Empire were over.

⇟ The Super Sci-Fi Heroes and Villains Quiz

1. **Let's start with a classic backstory—the orphaned superhero.
   Which superhero orphan was taken in by a nice couple named
   Jonathan and Martha?**

   Answer: Superman was born as Kal-El and shot into
   space by his real father to save him when their home
   planet, Krypton, was destroyed. The infant Kal-El landed
   in a Kansas cornfield and was adopted by Jonathan and
   Martha Kent, who named him Clark. That's where
   Clark learned his powerful Midwestern moral code that
   guides how he uses his superpowers.

2. **What color was not used as a color for kryptonite in any version of
   the Superman series?**

Answer: Radio scripts and comic book art portrayed the substance as red, grey, green, and metallic, but eventually the writers settled on green as the color of kryptonite.

3. **According to the comparison done by HowStuffWorks, who of the following would stand a chance of winning against Superman in a battle of wills?**

    Answer: D. While Superman can easily overpower most opponents, all three are particularly formidable foes and could use their superpowers to take advantage of Superman's few weaknesses.

4. **Which of the below is NOT part of the over-armor of Batman's awesome Batsuit?**

    Answer: The over-armor in the Batsuit is comprised of six pieces: knee guards, calf guards, leg armor, arm guards, a full-torso vest, and a spine guard.

5. **Which trait is NOT shared by Batman and Superman?**

    Answer: Both of these characters have secret identities and tragic family backstories, but Superman is the only one with real superpowers. Batman uses technology and training to simulate common hero abilities like flight and super strength.

6. **What is the sonic device that Batman has?**

Answer: The sonic device Batman carries in the heel of one of his boots can be used to summon swarms of bats instantly to create mass chaos at any scene. This allows Batman to create hellish diversions or make dramatic escapes.

7. **True or False: The Batmobile has a jet engine specifically designed to let it blast through walls or any other objects in its way.**

Answer: The Batmobile has a jet engine, yes, but it allows it to jump/fly through the air much farther than any normal car could. It's not specifically designed to blast through objects in its path.

8. **In the most accepted version of the Joker's origins, how does the Joker turn into the Joker?**

Answer: As revealed in *Detective Comics #168* (1951), the man who would become the Joker jumped into a vat of chemicals in order to escape from Batman. This origin is considered canon in the DC Comics community.

9. What is the major weapon on the superweapon, the Death Star, that the Sith created?

Answer: All three are parts of the Death Star's offensive and defensive systems, but the superlaser is the massive weapon that can destroy planets in a single blast.

10. True or False: Einstein's special theory of relativity makes it possible for the Starship *Enterprise* to use its superability, warp speed, to travel faster than the speed of light.

Answer: According to Einstein's special theory of relativity, nothing is faster than the speed of light; therefore space travel at warp speed would be impossible. It is Einstein's general theory of relativity that makes the concept of warp speed possible, because it allows for space and time to be bent and manipulated, thus avoiding the speed of light issue and making warp speed possible.

The Ultimate Superhero Quiz

1. After extensive consideration, who did HowStuffWorks conclude would win in a fight between Superman and Gandalf the magician?

Answer: HowStuffWorks concluded that because he's virtually immortal and able to conjure magic, the Lord of the Rings trilogy's Gandalf would eventually wear Superman down and win.

2. What is the name of the X-Men superhero who can absorb ordinary people's strength and memories as well as mutants' superpowers?

Answer: Rogue's mutant abilities allow her to absorb ordinary people's strength and memories, as well as mutants' superpowers.

3. Which one of the following criteria for antisocial personality disorder (APD) does Batman exhibit?

Answer: Batman uses heightened aggression with those he deems criminal, using both his body and external weapons to subdue his targets.

4. What are Wolverine's claws made of?

Answer: X-Man Wolverine's claws are made from adamantium. His skeleton consists of the same durable (and fictional) metal.

5. **What does the special memory fiber in Batman's cape allow him to do?**

Answer: Batman's cape can be sculpted into a kind of glider that can slow his falls or even allow for short flights.

6. **Which of the following *Star Trek* technologies actually has become a reality?**

Answer: Of the three, only hypospray—a form of hypodermic injection for medicine—has become a reality. In fact, it was one even before the creation of *Star Trek*. The real-world application is known as a jet injector.

7. **True or False: Salamanders are the most complex species of animals capable of regenerating body parts.**

Answer: Salamanders are the highest order of animals capable of regenerating body parts, including their tails, upper and lower jaws, eyes, and hearts. But hopefully human limb regeneration is on the horizon.

8. **True or False: The new superpower reality of liquid body armor is made from a Kevlar that has been soaked in a shear-thinning fluid to make give it a flexible, liquid layer and increased defensive strength.**

Answer: Liquid body armor is actually made from Kevlar that has been soaked in a shear-thickening fluid, which behaves like a solid when it encounters mechanical stress or shear. In other words, it moves like a liquid until an object strikes or agitates it forcefully. Then, it hardens in a few milliseconds.

9. **What is the main test to determine if someone is a replicant in Blade Runner?**

Answer: The Voigt-Kampff test used to determine who is a replicant measures involuntary biological reactions. First, the blade runner sets up his equipment, which monitors physical signals like muscle movement in the subject's eye. Once everything is ready and the subject sits still, the blade runner asks questions designed to touch on the subject's morals—questions which should produce a noticeable emotional response. The machine works a little like a lie detector, measuring subtle changes in a subject's blush response and eye motion.

10. **Which of the following technologies can be used to create an invisibility cloak?**

Answer: Carbon nanotubes, sheets of carbon wrapped

up into cylindrical tubes, are excellent conductors of heat, making them ideal mirage-makers. The carbon nanotube invisibility technology is based on the same idea as a mirage, only the technology creates the mirage itself, rather than detecting it.

SOURCES

Science Is Super Already: Top Five Science-Borne Superpowers

Anderson, James. "IBM Files Patent for Bullet Dodging Bionic Body Armor." *Techfragments*, February 13, 2009. Accessed November 6, 2009. www.techfragments.com/381/ibm-files-patent-for-bullet-dodging-bionic-body-armor/.

Atsma, Aaron. Theoi Greek Mythology: Exploring Mythology in Classical Literature & Art. Accessed November 6, 2009. www.theoi.com/Encyc_A.html.

Camero. "Xaver 800." Accessed November 6, 2009. www.camero-tech.com/product.php?ID=40.

CBC News. "Simon Fraser Engineers Create Gecko-Inspired Space Robots." January 2, 2014. Accessed November 24, 2014. www.cbc.ca/news/technology/simon-fraser-engineers-create-gecko-inspired-space-robots-1.2481912.

Digital Barriers. "ThruVision." Accessed November 12, 2009. www.digitalbarriers.com/products/thruvision/.

Ekso Bionics. Accessed November 6, 2009. www.eksobionics.com/#slide1.

Lockheed Martin. "HULC." Accessed November 6, 2009. www.lockheedmartin.com/us/products/hulc.html.

NanoRobotics Lab. "Gecko Hair Manufacturing." Accessed November 6, 2009. nanolab.me.cmu.edu/projects/geckohair/.

North American Weather Consultants Inc. "Cloud Seeding Frequently

Asked Questions." Accessed November 6, 2009. www
.nawcinc.com/wmfaq.html.

Weather Modification Association. "Position Statement on the
Environmental Impact of Using Silver Iodide as a Cloud
Seeding Agent." July 2009. Accessed November 6, 2009.
www.weathermodification.org/images/AGI_toxicity.pdf.

Wilson, Tracy V. "How Liquid Body Armor Works." HowStuffWorks
.com. Accessed November 6, 2009. science.howstuffworks
.com/liquid-body-armor1.htm.

Yang, Sarah. "Engineers Create New Adhesive That Mimics Gecko
Toe Hairs." UC Berkeley News. January 29, 2008. Accessed
November 6, 2009. www.berkeley.edu/news/media
/releases/2008/01/29_gecko.shtml.

* * *

⚡ How Comic Books Work

Arnold, Andrew. "Blockbuster Summer: A Short Comic-Book History."
Time, May 20, 2002. Accessed May 19, 2011. www.time.com
/time/magazine/article/0,9171,1002488,00.html.

Bluewater Productions. "Comics." Accessed May 19, 2011. www
.bluewaterprod.com/comics/comics.php.

Box Office Mojo. "Comic Book Adaptation." Accessed May 19, 2011.
boxofficemojo.com/genres/chart/?id=comicbookadaptation
.htm.

The Comic Book Database. "Top Rated Issues." Accessed May 19, 2011.
www.comicbookdb.com/top_ratings.php.

The Comic Books. "Newsstand Period: 1922–1955." Accessed May 19,
2011. www.thecomicbooks.com/nsp1-01.html.

Contino, Jennifer M. "An Astonishing Colorist: Laura Martin."
Sequential Tart, March 1, 2006. Accessed May 19. 2011. www
.sequentialtart.com/article.php?id=68.

Eldred, Tim. "Manga and Comic Book History." Cosmo DNA. June 25, 2013. Accessed May 19, 2011. ourstarblazers.com/vault/170/.

Elfring, Mat. "Comic Book History: Comics Code Authority." *Comic Vine*, June 7, 2010. Accessed May 19, 2011. www.comicvine.com /news/comic-book-history-comics-code-authority/141329/.

Flood, Alison. "Donald Trump Comic Book Tells Tycoon's Life Story." *The Guardian*, May 11, 2011. Accessed May 19, 2011. www.guardian .co.uk/books/2011/may/11/donald-trump-comic-book.

Glazer, Sarah. "Manga for Girls." *New York Times*, September 18, 2005. Accessed May 19, 2011. www.nytimes.com/2005/09/18 /books/review/18glazer.html.

Grand Comics Database. "Detective Comics." Accessed May 19, 2011. www.comics.org/series/87/.

Lawson, Terry. "Director Brings Vampire Comic Books to Life." *Bismarck Tribune*, December 9, 2004. Accessed May 19, 2011. www.bismarcktribune.com/news/local/article_21e6288e -6af5-5398-8eee-eccbc953bbc0.html.

Martin, Laura (comic book colorist), personal interview with author, May 19, 2011.

Masters, Coco. "America Is Drawn to Manga." *Time*, August 10, 2006. Accessed May 19, 2011. www.time.com/time/magazine /article/0,9171,1223355,00.html.

Matheson, Whitney. "Free Comic Day Preview: Ten Comics You Should Grab." *USA Today*, May 6, 2011. Accessed May 19, 2011. content.usatoday.com/communities/popcandy/post/2011/05 /free-comic-book-day-preview-my-top-10-comics/1.

McCloud, Scott. "Infinite Canvas." Scott McCloud's website. February 2009. Accessed May 19, 2011. www.scottmccloud.com/4 -inventions/canvas/index.html.

McCormack, Claire. "Can a Disabled Muslim Comic-Book Superhero Save the World?" *Time*, March 8, 2011. Accessed May 19, 2011. www.time.com/time/arts/article/0,8599,2048930,00.html.

Molotiu, Andrei. "History of Comic Book Art." Reading with Pictures. Accessed May 19, 2011. www.readingwithpictures.org/wp -content/uploads/2008/03/History-of-Comic-Book-Art-by -Prof-Andrei-Molotiu.pdf.

Nostomania. "Nostomania's 100 Most Valuable Comic Books." Accessed May 19, 2011. www.nostomania.com/servlets/com .nostomania.CatPage?name=Top100ComicsMain.

Pink, Daniel H. "Japan, Ink: Inside the Manga Industrial Complex." *Wired*, October 22, 2007. Accessed May 19, 2011. archive.wired. com/techbiz/media/magazine/15-11/ff_manga?currentPage =all.

Quattro, Ken. "The New Ages: Rethinking Comic Book History." Comicartville. 2004. Accessed May 19, 2011. www .comicartville.com/newages.htm.

Siklos, Richard. "Spoiler Alert: Comic Books Are Alive and Kicking." *Fortune*, October 13, 2008. Accessed May 19, 2011. money.cnn .com/2008/10/10/news/companies/siklos_marvel.fortune/.

St. Mary, Rob. "Industry Troubles? Free Comic Books to the Rescue!" NPR. May 6, 2011. Accessed May 19, 2011. www.npr .org/2011/05/06/136051637/industry-troubles-free-comic -books-to-the-rescue.

Stelfreeze, Brian (comic-book cover artist), personal interview with author, May 17, 2011.

Stevens, Tim. "SDCC 2009: Laura Martin Goes Exclusive." Marvel Comic News. July 31, 2009. Accessed May 19, 2011. marvel .com/news/story/8923/sdcc_2009_laura_martin_goes_exclusive.

University of Missouri Library, Special Collections and Rare Books. "Comic Art Collection." Accessed May 19, 2011. mulibraries .missouri.edu/specialcollections/comic.htm.

Van Gelder, Lawrence. "A German Comic Book on Holocaust History." *New York Times*, February 4, 2008. Accessed May

19, 2011. www.nytimes.com/2008/02/04/arts/04arts-AGER
    MANCOMIC_BRF.html.

Wehrum, Kasey. "Comic Books for Entrepreneurs." *Inc.*, May 2011.
    Accessed May 19, 2011. www.inc.com/magazine/20110501
    /comic-books-for-entrepreneurs.html.

Weldon, Glen. "Amid Slack Sales, the Comics Industry Once Again
    Tightens Its Utility Belt." NPR. April 14, 2011.
    Accessed May 19, 2011. www.npr.org/blogs/monkeysee
    /2011/04/14/135381019/amid-slack-sales-the-comics-in
    dustry-once-again-tightens-its-utility-belt.

Weldon, Glen. "Censors and Sensibility: RIP, Comics Code
    Authority Seal of Approval, 1954-2011." NPR. January
    27, 2011. Accessed May 19, 2011. www.npr.org/blogs
    /monkeysee/2011/01/27/133253953/censors-and-sensibility
    -rip-comics-code-authority-seal-1954-2011.

Weldon, Glen. "The Week in Comics: Palicki Is Prince, Spidey
    Gets Nubbly, Comics Are Doomed." NPR. February
    17, 2011. Accessed May 19, 2011. www.npr.org/blogs
    /monkeysee/2011/02/23/133834049/the-week-in-comics
    -palicki-is-prince-spidey-gets-nubbly-comics-are-doomed.

Wood, Mary. "The Yellow Kid on the Paper Stage." University of
    Virginia, American Studies. February 2, 2004. Accessed May
    19, 2011. xroads.virginia.edu/~MA04/wood/ykid/yj.htm.

* * *

Heroes Unmasked: How Secret Are Secret Identities

No sources.

* * *

⚡ How Kryptonite Works

Alfred, Mark. "The Colors Out of Space." Superman Thru the Ages. Accessed June 19, 2006. supermanthrutheages.com/articles /space-colors/.

Beatty, Scott. *Superman: The Ultimate Guide to the Man of Steel.* New York: DK Books, 2002.

Goulart, Ron. *Great History of Comic Books.* Chicago: Contemporary Books, 1986.

National Aeronautics and Space Administration. "NASA's Imagine the Universe: X-Ray Detectors." Accessed June 19, 2006. imagine .gsfc.nasa.gov/docs/science/how_l2/xray_detectors.html.

Rozakis, Bob and John Wells. "Kryptonite: Part One." Silver Bullet Comic Books.

Rozakis, Bob and John Wells. "Kryptonite: Part Two." Silver Bullet Comic Books.

Rozakis, Bob and John Wells. "Kryptonite: Part Three." Silver Bullet Comic Books.

Seeds, Michael. *Foundations of Astronomy.* Stamford, CT: Wadsworth Publishing, 1994.

Superman Home Page. "DC Comics Q&A." Accessed June 19, 2006. www.supermanhomepage.com/comics/comics.php?topic =comics-racfaq#Kryptonite.

* * *

⚡ Battling Blockbusters: Fighting against Superman

Beatty, Scott. *Superman: The Ultimate Guide to the Man of Steel.* New York: DK Books, 2002.

Fisher, Mark. *The Encyclopedia of Arda.* Accessed June 23, 2006. www .glyphweb.com/arda/.

Marvel Directory. "Galactus." Accessed June 23, 2006. www .marveldirectory.com/individuals/g/galactus.htm.

Superman Home Page. "Who's Who in the Superman Comics: Doomsday." Accessed June 23, 2006. www.supermanhome page.com/comics/who/who-intro.php?topic=doomsday.

Uncanny X-Men.net. "Rogue." Accessed June 23, 2006. www .uncannyxmen.net/db/spotlight/showquestion.asp?faq=10&fld Auto=56.

* * *

The Dark Knight in Shining Armor: How Batman and His Batsuit Work

Internet Movie Database. "*Batman Begins.*" Accessed March 11, 2013. www.imdb.com/title/tt0372784.

Warner Bros. "*Batman Begins.*" Accessed March 11, 2013. www2 .warnerbros.com/batmanbegins/index.html.

* * *

No Laughing Matter: How the Joker Works

No sources.

* * *

The Man Behind the Mask: Is Batman a Sociopath?

No sources.

* * *

May the Force Be with You: How Does a Star Wars Lightsaber Work?

No sources.

* * *

To the Batmobile, Robin! How the Batmobile Works

Crowley, Nathan (Batmobile designer), personal interview with author, June 2005.

* * *

Darth Vader's Driving Force: The Sith Explained

Anderson, Kevin J. *Jedi Search (Star Wars: The Jedi Academy Trilogy, Volume 1)*. New York: Bantam Books, 1994.

Anderson, Kevin J., et al. *Tales of the Jedi: The Golden Age of the Sith*. Dark Horse Comics, 1997.

Anderson, Kevin J., et al. *Tales of the Jedi: The Sith War*. Dark Horse Comics, 1997.

Bouzereau, Laurent. *Star Wars: The Annotated Screenplays*. New York: Del Rey Books, 1997.

Bouzereau, Laurent and Jody Duncan. *The Making of Star Wars, Episode I: The Phantom Menace*. New York: Random House, 1999.

Brooks, Terry. *Star Wars, Episode I: The Phantom Menace*. New York: Del Rey Books, 1999.

Lucasfilm. *Star Wars Encyclopedia*. Accessed May 18, 2005. starwars.com /explore/encyclopedia.

Lucas, George, et al. *Star Wars I: The Phantom Menace*. DVD. Directed by George Lucas. CTV Services, Tunisia: Twentieth Century Fox, 2005.

Lucas, George, et al. *Star Wars II: The Attack of the Clones*. DVD. Directed by George Lucas. Chott el Djerid, Nefta, Tunisia: Twentieth Century Fox, 2005.

Lucas, George, et al. *Star Wars III: The Revenge of the Sith*. DVD. Directed by George Lucas. Chott el Djerid, Nefta, Tunisia: Twentieth Century Fox, 2005.

Lucas, George, et al. *Star Wars IV: A New Hope*. DVD. Directed by George Lucas. Durango, Mexico: Twentieth Century Fox, 2006.

Lucas, George, et al. *Star Wars V: The Empire Strikes Back*. DVD. Directed by George Lucas. Banks, Oregon: Twentieth Century Fox, 2006.

Lucas, George, et al. *Star Wars VI: Return of the Jedi*. DVD. Directed by George Lucas. Buttercup Valley, California: Twentieth Century Fox, 2006.

Reynolds, David West. *Star Wars Episode I: Visual Dictionary*. New York: DK Books, 1999.

Reynolds, David West, Hans Jenssen, and Richard Chasemore. *Incredible Cross-Sections of Star Wars Episode I: The Definitive Guide to the Craft*. New York: DK Books, 1999.

Sansweet, Stephen J. *Star Wars Encyclopedia*. New York: Del Rey Books, 1998.

Stackpole, Michael A. *I, Jedi: Star Wars*. New York: Bantam Books, 1999.

Star Wars Behind the Magic: The Insider's Guide to Star Wars. Los Angeles, CA: LucasArts. CD-ROM.

Star Wars Episode I: Insider's Guide. Los Angeles, CA: LucasArts, 1999. CD-ROM.

Star Wars Jedi Knight: Jedi Academy. Los Angeles, CA: LucasArts, 2003. Video game.

Star Wars Jedi Knight: Mysteries of the Sith. Los Angeles, CA: LucasArts, 1998. Video game.

Star Wars Jedi Outcast: Jedi Academy. Los Angeles, CA: LucasArts, 2002. Video game.

Star Wars Knights of the Old Republic. Los Angeles, CA: LucasArts, 2003. Video game.

Star Wars Knights of the Old Republic 2: The Sith Lords. Los Angeles, CA: LucasArts, 2005. Video game.

Star Wars Timeline, a display at the Denver Star Wars Celebration, April 30-May 2, 1999.

Topps Widevision Cards. *Star Wars I: The Phantom Menace*. Trading cards.

Tyers, Kathy. *The Truce at Bakura: Star Wars*. New York: Bantam Books, 1994.

Wallace, Daniel. *Star Wars Episode I, What's What: A Pocket Guide to The Phantom Menace*. Philadelphia: Running Press, 1999.

Windham, Ryder. *Star Wars Episode I: The Phantom Menace Movie Scrapbook*. New York: Random House, 1999.

Zahn, Timothy. *Heir to the Empire (Star Wars: The Thrawn Trilogy, Volume 1)*. New York: Bantam Books, 1993.

Zahn, Timothy. *The Last Command (Star Wars: The Thrawn Trilogy, Volume 3)*. New York: Bantam Books, 1994.

* * *

⚡ Darth Vader's Driving Force: The Sith Explained

No sources.

* * *

⚡ Controlling Chaos: How the Death Star Works

Hughes, Sam. "How to Destroy the Earth." Things of Interest. Accessed November 1, 2005. qntm.org/destroy.

Jedi Council Forums. "A History of the Death Stars: Another TalonCard Project." Accessed November 1, 2005. boards.theforce .net/threads/a-history-of-the-death-stars-another-taloncard -project-tm.13120474/page-14.

NASA Jet Propulsion Laboratory . "PIA05423: That's No Space Station." July 26, 2004. photojournal.jpl.nasa.gov/catalog/PIA05423.

Saxton, Curtis. "Death Stars: Construction and Destruction." Star Wars Technical Commentaries. Accessed November 1, 2005. www .theforce.net/swtc/ds/.

Star Wars Encyclopedia. "Death Star." starwars.com/explore /encyclopedia/technology/deathstar/?id=eu.

Star Wars Encyclopedia. "Galactic Empire." starwars.com/explore /encyclopedia/groups/empire/.

Star Wars Encyclopedia. "Tarkin." starwars.com/explore/encyclopedia /characters/tarkin/.

University of Georgia, Department of Physics and Astronomy. "Ask The Physicist." www.physast.uga.edu/ask_phys.html.

Wong, Michael. "Death Star Firepower." Star Destroyer. July 24, 1999. www.stardestroyer.net/Empire/Tech/Beam/DeathStar.html.

Wookiepedia. "Cassio Tagge." Accessed November 1, 2005. starwars .wikia.com/wiki/Cassio_Tagge.

Wookiepedia. "Superlaser." Accessed November 1, 2005. starwars.wikia .com/wiki/Superlaser.

Wookiepedia. "Tiann Jerjerrod." Accessed November 1, 2005. starwars .wikia.com/wiki/Tiaan_Jerjerrod.

* * *

⚡ All Aboard the Starship Enterprise: How Star Trek's Warp Speed Works

No sources.

* * *

⚡ Beam Me Up, Scotty: How Teleportation Will Work

No sources.

* * *

⚡ Tried and True Trekkie Tech: Top Ten *Star Trek* Technologies That Actually Came True

Applied Energetics. "Laser Guided Energy." Accessed November 9, 2009. ionatron.net/laser-guided-energy.asp.

AT&T. "AT&T First Service Provider to Deliver Intercompany Cisco Telepresence for Business around the World." April 21, 2008. Accessed October 21, 2009. www.att.com/gen/press-room?pid=4800&cdvn=news&newsarticleid=25523.

Bartkewicz, Anthony. "Company Creates Star Trek Coffins." KRQE TV, Albuquerque. April 6, 2009. Accessed October 21, 2009. www.krqe.com/dpp/news/strange/offbeat_dpgo_Company _creates_Star_Trek_coffins_SAB_040420092296082.

Batchelor, David Allen. "The Science of Star Trek." National Aeronautics and Space Administration. Accessed October 22, 2009. www .nasa.gov/topics/technology/features/star_trek.html.

BBC News. "'Star Trek Device' Could Detect Illness." September 20, 2002. Accessed October 20, 2009. ncws.bbc.co.uk/2/hi/health /2231989.stm.

Cisco. "Telepresence." Accessed October 21, 2009. www.cisco.com /en/US/products/ps7060/index.html#,hide-id-trigger-g1 -room_environments.

Coulter, Dauna. "Space Station Tricorder." Science@NASA. May 9, 2008. Accessed November 6, 2009. science.nasa.gov/headlines /y2008/09may_tricorder.htm?list185546.

Cruz, Gilbert. "Jack Cover." *Time*, February 19, 2009. Accessed October 19, 2009. www.time.com/time/magazine/article /0,9171,1880636,00.html.

Lundin, Laura. "Air Force Testing New Transparent Armor." U.S. Air Force. October 17, 2005. Accessed October 21, 2009. www .freerepublic.com/focus/f-news/1504430/posts.

Mick, Jason. "New 'Miracle Diagnosis' Handheld Medical Scanner 800 Times More Sensitive Than Full-Size Scanners." *Daily Tech*, July 10, 2008. Accessed October 21, 2009. www.dailytech .com/article.aspx?newsid=12322.

National Institute of Standards and Technology. "'Femtomolar Optical

Tweezers' May Enable Sensitive Blood Tests." Accessed October 20, 2009. www.nist.gov/pml/div684/tweezers_111208.cfm.

Schirber, Michael. "Doctors Could Go Needle-Free, But Sticking Points Remain." *Live Science*, October 4, 2006. Accessed October 21, 2009. www.livescience.com/health/061004_needle_free.html.

Stanford University. "Optical Tweezers: An Introduction." Accessed October 22, 2009. www.stanford.edu/group/blocklab/Optical %20Tweezers%20Introduction.htm.

Star Trek Database. "Geordi La Forge." Accessed October 20, 2009. www.startrek.com/database_article/la-forge-Geordi.

Vocera. Accessed October 22, 2009. www.vocera.com/products.

Young, Kelly. "'Bionic Eye' May Help Reverse Blindness." *New Scientist*, March 31, 2005. Accessed October 22, 2009. www .newscientist.com/article/dn7216.

★ ★ ★

⚡ No More Phantom Limbs: How Can Salamanders Regrow Body Parts?

Brockes, Jeremy P. and Anoop Kumar. "Appendage Regeneration in Adult Vertebrates and Implications for Regenerative Medicine." *Science*, December 23, 2005. Accessed October 28, 2008. www .sciencemag.org/cgi/content/full/310/5756/1919.

Bryner, Jeanna. "How Salamanders Sprout New Limbs." *LiveScience*, November 1, 2007. Accessed October 28, 2008. www .livescience.com/animals/071101-newt-limbs.html.

Kotulak, Ronald. "Research Brings Hope Body Parts Can Regrow." *Chicago Tribune*, September 29, 2006. articles.chicagotribune .com/2006-09-29/news/0609290295_1_genes-scars-body-parts.

Kumar, Anoop, et al. "Molecular Basis for the Nerve Dependence of Limb Regeneration in an Adult Vertebrate." *Science*, November

2, 2007. Accessed October 28, 2008. www.sciencemag.org /cgi/content/full/318/5851/772/DC2.

Muneoka, Ken, Manjong Han, and David M. Gardiner. "Regrowing Human Limbs." *Scientific American*, March 17, 2008. Accessed October 28, 2008. www.sciam.com/article.cfm?id=regrowing -human-limbs.

Philipkoski, Kristen. "Grow Your Own Limbs." *Wired*, September 22, 2006. Accessed October 28, 2008. archive.wired.com /medtech/genetics/news/2006/09/71817?currentPage=all.

Ritter, Malcolm. "Regrowing of Fingers Gets Set for Trial Stage." *St. Louis Post-Dispatch*, February 19, 2007.

Stocum, David L. *Regenerative Biology and Medicine*. Burlington, MA: Academic Press, 2006.

Tsonis, Panagiotis. "Limb Regeneration." Cambridge, UK: Cambridge University Press, 1996.

* * *

How the Future Force Warrior Will Work

No sources.

* * *

I'm Invincible! How Body Armor Works

No sources.

* * *

Fluid Fortification: How Liquid Body Armor Works

Arndt, Michael. "Body Armor Fit for a Superhero." *Business Week*, August 6, 2006. Accessed January 26, 2007. www.businessweek.com /magazine/content/06_32/b3996068.htm.

Baard, Erik. "Space-Age Goop Morphs between Liquid and Solid." Space.com. September 5, 2001. Accessed January 26, 2007. archive.today/mqC90.

Gladek, Eva. "Liquid Armor." ScienCentral News. June 15, 2006. Accessed January 26, 2007. www.sciencentral.com/articles /view.php3?type=article &article_id=218392807.

Johnson, Tonya. "ARL Scientists and Engineers Develop Liquid Armor Based on Nanotechnology." *RDECOM*, February 2004.

Johnson, Tonya. "Army Scientists, Engineers Develop Liquid Body Armor." Military.com. April 21, 2004. Accessed January 26, 2007. www .military.com/NewsContent/0,13319,usa3_042104.00.html.

Lee, Y.S. et al. "Advanced Body Armor Utilizing Shear Thickening Fluids." University of Delaware. Accessed January 26, 2007. www.ccm.udel.edu/STF/PubLinks2/AdvancedBodyArmor _Pres.pdf.

LORD Corporation. "Coatings." Accessed January 26, 2007. www.lord .com/products-and-solutions/coatings.xml.

Love, Lonnie J. "Ferrofluid." AccessScience. October 27, 2006. Accessed January 27, 2007. www.accessscience.com/content /ferrofluid/801330.

Lurie, Karen. "Instant Armor." ScienCentral News. December 4, 2003. Accessed January 27, 2007. www.sciencentral.com/articles /view.php3? article_id=218392121&language=english.

Markovitz, Hershel. "Rheology." AccessScience. August 26, 2005. Accessed January 27, 2007. www.accessscience.com/content /rheology/586100.

University of Delaware. "Shear-Thickening Fluid." Accessed January 27, 2007. www.ccm.udel.edu/STF/pubs1.html.

Weist, John M. "Non-Newtonian Fluid." AccessScience. August 25, 2005. Accessed January 27, 2007. www.accessscience.com /content/non-newtonian-fluid/455810.

Wetzel, Eric D., et al. "Advanced Body Armor Utilizing Shear

Thickening Fluids." Presentation at Army Science Conference on Composite Materials Research, Orlando, FL, December 3, 2002. Accessed January 26, 2007. www.ccm.udel.edu/STF /PubLinks2/AdvancedBodyArmor_Pres.pdf.

Wetzel, Eric D., et al. "'Liquid Armor': Protective Fabrics Utilizing Shear Thickening Fluids." Presentation at Industrial Fabrics Association International's Conference on Safety and Protective Fabrics, Pittsburgh, PA, October 27, 2004. Accessed January 26, 2007. www.ccm.udel.edu/STF/PubLinks2 /LiquidArmorProtectiveFabrics_Pub.pdf.

* * *

Invasion of the Humanoids: How Replicants Work

Brian Kelly, et al. *Blade Runner*. DVD. Directed by Ridley Scott. Burbank, California: Twentieth Century Fox, 1982.

Dick, Philip K. *Do Androids Dream of Electric Sheep?* New York: Del Rey Books, 1996.

Greenwald, Ted. "Q&A: Ridley Scott Has Finally Created the 'Blade Runner' He Always Imagined." *Wired*, September 26, 2007. Accessed October 1, 2007. archive.wired.com/entertainment /hollywood/magazine/15-10/ff_bladerunner?currentPage=all.

Kaplan, Fred. "A Cult Classic Restored, Again." *New York Times*, September 30, 2007. Accessed October 1, 2007. www.nytimes .com/2007/09/30/movies/30kapl.html.

Mori, Masahiro. "The Uncanny Valley." *Energy* 7 (1970): 33-35. Accessed October 18, 2007. www.androidscience.com /theuncannyvalley/proceedings2005/uncannyvalley.html.

Philip K. Dick: The Official Site. "Films." Accessed October 18, 2007.

Whitehouse, David. "Japanese Develop 'Female' Android." BBC News. July 27, 2005. Accessed October 15, 2007. news.bbc.co.uk/2 /hi/science/nature/4714135.stm.

* * *

⋛ Top Five Sci-Fi Weapons That Might Actually Happen

Bozkurt, Alpert, Amit Lal, and Robert F. Gilmour. "Electrical Endogenous Heating of Insect Muscles for Flight Control." Presentation at the 30th International Conference of IEEE Engineering in Medicine and Biology Society, Vancouver, Canada, August 20, 2008. Accessed November 5, 2009. www2.lirmm.fr/lirmm /interne/BIBLI/CDROM/ROB/2008/EMBC_2008/PDFs c./Papers/14560054.pdf.

Bozkurt, Alpert, Amit Lal, and Robert F. Gilmour. "Radio Control of Insects for Biobotic Domestication." Presentation at Fourth International IEEE/EMBS Conference on Neural Engineering, April 29 2009-May 2, 2009. Accessed November 5, 2009. ieeexplore.ieee.org/xpl/login.jsp?tp=&arn umber=5109272&url=http%3A%2F%2Fieeexplore.ieee.org %2Fstamp%2Fstamp.jsp%3Ftp%3D%26arnumber%3D5109272.

Bozkurt, Alpert, Robert F. Gilmour, and Amit Lal. "Balloon-Assisted Flight of Radio-Controlled Insect Biobots." *IEEE Transactions on Biomedical Engineering* 56 (September 2009): 2304–2307. bio instrumentacion.eia.edu.co/Documentacion/Bio/insectos.pdf.

Jewell, Mark. "Robotic Suit Could Usher in Super Solider Era." NBC News. May 15, 2008. Accessed November 5, 2009. www.msnbc .msn.com/id/24651455/.

Morrison, David. "FAQs about NEO Impacts." NASA Ames Research Center, Asteroid and Comet Hazards. September 2004. Accessed November 5, 2009. impact.arc.nasa.gov/intro_faq.cfm.

Morrison, David. "Introduction." NASA Ames Research Center, Asteroid and Comet Hazards. September 2004. Accessed November 5, 2009. impact.arc.nasa.gov/intro.cfm.

Nave, C.R. "Hafele and Keating Experiment." HyperPhysics, Georgia State University Department of Physics and Astronomy.

Accessed November 2, 2009. hyperphysics.phy-astr.gsu.edu /HBASE/relativ/airtim.html.

Plait, Phil. "The Astronomy of Armageddon." *Bad Astronomy*, December 28, 2008. Accessed November 5, 2009. www.badastronomy .com/bad/movies/armpitageddon.html.

Vergano, Dan. "Real Spying Squirrels, Dolphins Helped Inspire 'G-Force.'" *USA Today*, July 27, 2009. Accessed November 5, 2009. www.usatoday.com/tech/science/columnist/vergano /2009-07-25-g-force_N.htm.

Yeomans, Donald K., et al. "Deflecting a Hazardous Near-Earth Object." Presentation at the First IAA Planetary Defense Conference: Protecting Earth from Asteroids, Granada, Spain, April 27–30, 2009. neo.jpl.nasa.gov/neo/pdc_paper.html.

* * *

Now You See Me, Now You Don't: How Invisibility Cloaks Work

Adler, Robert. "Acoustic 'Superlens' Could Mean Finer Ultrasound Scans." *New Scientist*, January 8, 2008. Accessed October 13, 2009. www.newscientist.com/article/dn13156-acoustic-super lens-could-mean-finer-ultrasound-scans.html.

Aliev, Ali E., et al. "Mirage Effect from Thermally Modulated Transparent Carbon Nanotube Sheets." *Nanotechnology* 22 (October 28, 2011). Accessed October 13, 2011. nanotech.utdallas.edu /documents/Mirage.pdf.

Barras, Colin. "Gold Rings Create First True Invisibility Cloak." *New Scientist*, October 2, 2007. Accessed October 13, 2009. www .newscientist.com/article/dn12722-gold-rings-create-first -true-invisibility-cloak.html.

BBC News. "Inventor Plans 'Invisible Walls.'" June 14, 2004. news.bbc .co.uk/2/hi/technology/3791795.stm.

Bland, Eric. "Invisibility Cloak Closer Than Ever to Reality." Discovery News. January 15, 2009. Accessed October 13, 2009. dsc .discovery.com/news/2009/01/15/invisibility-cloak.html.

Brown, Mark. "Watch: 'Invisibility Cloak' Uses Mirages to Make Objects Vanish." *Wired*, October 4, 2011. Accessed October 13, 2011. www.wired.com/dangerroom/2011/10/invisibility-cloak-mirage/.

Duke University, Pratt School of Engineering. "'Invisibility Cloaks' Could Break Sound Barriers." January 9, 2008. Accessed October 13, 2009. www.pratt.duke.edu/news /%C3%A2%E2%82%AC%CB%9Cinvisibility-cloaks%C3%A 2%E2%82%AC%E2%84%A2-could-break-sound-barriers.

Feiner, Steven K. "Augmented Reality: A New Way of Seeing." *Scientific American*, April 2002. www.scientificamerican.com/article .cfm?id=augmented-reality-a-new-w.

Inami, Masahiko, et al. "Optical Camouflage Using Retro-Reflective Projection Technology." Proceedings of the Second IEEE and ACM International Symposium on Mixed and Augmented Reality (ISMAR 03), Tokyo, Japan, October 7-10, 2003. Accessed October 13, 2009. files.tachilab.org/publications /intconf2000/inami2003ISMAR.pdf.

Inami, Masahiko, et al. "Visuo-Haptic Display Using Head-Mounted Projector." University of Tokyo, March 18-22, 2000. Accessed October 13, 2009. star.t.u-tokyo.ac.jp/projects/MEDIA/xv /vr2000.pdf.

McCarthy, Wil. "Being Invisible." *Wired*, November 2008. Accessed October 13, 2009. www.wired.com/wired/archive/11.08 /pwr_invisible_pr.html.

Mullins, Justin. "Working Invisibility Cloak Created at Last." *New Scientist*, October 19, 2006. Accessed October 13, 2009. www .newscientist.com/article/dn10334-working-invisibility-cloak -created-at-last.html.

Pendry, John. "Metamaterials." *New Scientist, Instant Expert* 7. Accessed

October 21, 2011. www.newscientist.com/data/doc/article
/dn19554/instant_expert_7_-_metamaterials.pdf.

Smolyaninov, Igor, et al. "Electromagnetic Cloaking in the Visible
Frequency Range." University of Maryland, Department
of Electrical and Computer Engineering. December 10,
2007. Accessed October 13, 2009. arxiv.org/ftp/arxiv
/papers/0709/0709.2862.pdf.

Tachi, Susumu, et al. "Telexistence and Retro-Reflective Projection
Technology," Proceedings of the Fifth Virtual Reality
International Conference (VRIC2003), Laval, France, May
14-16, 2003. Accessed October 21, 2011. citeseerx.ist.psu.edu
/viewdoc/summary?doi=10.1.1.102.8219.

* * *

⚡ Wishful Thinking: Ten Sci-Fi Gadgets We Wish Actually Existed

Adams, Douglas. More Than Complete Hitchhiker's Guide: Complete &
Unabridged. Stamford, CT: Longmeadow, 1987.

BBC. "Doctor Who." Accessed September 22, 2011. www.bbc.co.uk
/doctorwho/dw.

Drexler, K. Eric, et al. "Debate about Assemblers: Smalley Rebuttal."
Institute for Molecular Manufacturing. 2001. Accessed
September 22, 2011. www.imm.org/publications/sciam
debate2/smalley/.

Marvel Universe Wiki. "Iron Man (Anthony Stark)." Accessed September
22, 2011. marvel.com/universe/Iron_Man_(Anthony_Stark).

Smalley, Richard E. "Of Chemistry, Love, and Nanobots." Scientific
American, September 2001. Accessed October 5, 2011.
cohesion.rice.edu/NaturalSciences/Smalley/emplibrary/SA2
85-76.pdf.

Sonnenfeld, Barry, et al. Men in Black. DVD. Directed by Barry

Sonnenfeld. Universal City, California: Amblin Entertainment, 1997.

StarTrek.com. Accessed September 22, 2011. www.startrek.com/.

StarWars.com. Accessed September 22, 2011. www.starwars.com/.

Zemeckis, Robert, et al. *Back to the Future*. DVD. Directed by Robert Zemeckis. Universal City, California: Universal Pictures, 1985.

Zemeckis, Robert, et al. *Back to the Future, Part II*. DVD. Directed by Robert Zemeckis. Universal City, California: Universal Pictures, 1987.

CONTRIBUTORS

Kevin Bonsor

Marshall Brain

Josh Briggs

Nathan Chandler

Cristen Conger

John Fuller

Tom Harris

Matt Hunt

Chris Jones

Susan L. Nasr

Chris Pollette

Jacob Silverman

Jonathan Strickland

Robert Valdes

Tracy V. Wilson

ABOUT HOWSTUFFWORKS

HowStuffWorks.com is an award-winning digital source of credible, unbiased, and easy-to-understand explanations of how the world actually works. Founded in 1998, the site is now an online resource for millions of people of all ages. From car engines to search engines, from cell phones to stem cells, and thousands of subjects in between, HowStuffWorks.com has it covered. In addition to comprehensive articles, our helpful graphics and informative videos walk you through every topic clearly and objectively. Our premise is simple: demystify the world and do it in a clear-cut way that anyone can understand.

If you enjoyed *The Science of Superheroes and Space Warriors: Lightsabers, Batmobiles, Kryptonite, and More!*, check out the rest of the series from Sourcebooks and HowStuffWorks.com!

THE REAL SCIENCE OF SEX APPEAL: WHY WE LOVE, LUST, AND LONG FOR EACH OTHER

EVER WONDER WHY LOVE MAKES US SO CRAZY? COME DIVE INTO THE *REAL* SCIENCE BEHIND SEX APPEAL AND WHY WE LOVE, LUST, AND LONG FOR EACH OTHER.

Did you know your walk, your scent, and even the food you eat can make you sexier? Or that there are scientifically proven ways to become more successful at dating, especially online? The team at the award-winning website HowStuffWorks reveals the steamy science of love and sex, from flirting to falling in love and everything in between. Discover:

* How aphrodisiacs and sex appeal work (and how to increase yours!)

* Whether love at first sight is scientifically possible

* Why breakup songs hurt so good

* What happens in the brain during an orgasm

* The crazy chemistry behind long-term relationships

* The dope on dating and matchmaking

* And much more!

This dynamic book will show you what to expect—and what to do—the next time someone sets your heart racing.

FUTURE TECH, RIGHT NOW: X-RAY VISION, MIND CONTROL, AND OTHER AMAZING STUFF FROM TOMORROW

FROM X-RAY VISION TO MIND READING, THE FUTURE IS COMING ON FAST!

Flying cars! Teleporting! Robot servants! Wouldn't you love any of these? You're in luck because they may be closer to reality than you think. Based on the best of HowStuffWorks' popular podcasts *TechStuff*, *Stuff from the Future*, and *Stuff to Blow Your Mind*, this dynamic book reveals the science of our future, from mind control and drugs that can make you smarter to textbooks that talk to you and even robotic teammates. Discover:

* How telekinesis and digital immortality work

* Whether computers could replace doctors one day

* What robot servants and coworkers will look like

* Five of the coolest future car technologies

* What we will do for fun in 2050

* And much more!

Come explore the coolest and craziest technology of the future.